Lab Manual

ELECTRICITY, ELECTRONICS, AND CONTROL SYSTEMS FOR HVAC

FOURTH EDITION

Thomas E. Kissell
Terra Community College

Upper Saddle River, New Jersey
Columbus, Ohio

Editor in Chief: Vernon Anthony
Editor: Eric Krassow
Associate Managing Editor: Christine Buckendahl
Editorial Assistant: Nancy Kesterson
Production Coordination: Naomi Sysak
Project Manager: Holly Shufeldt
Design Coordinator: Diane Ernsberger
Operations Specialist: Laura Weaver
Director of Marketing: David Gesell
Executive Marketing Manager: Derril Trakalo
Marketing Assistant: Les Roberts

This book was set in Times by Naomi Sysak. It was printed and bound by Bind-Rite Graphics. The cover was printed by Phoenix Color Corp.

Pearson Education Ltd.
Pearson Education Singapore Pte. Ltd.
Pearson Education Canada, Ltd.
Pearson Education—Japan
Pearson Education Australia Pty. Limited
Pearson Education North Asia Ltd.
Pearson Educación de Mexico, S.A. de C.V.
Pearson Education Malaysia Pte. Ltd.

10 9 8 7 6 5 4 3 2 1
ISBN-13: 978-0-13-199570-3
ISBN-10: 0-13-199570-7

CONTENTS

Preface

This lab manual is designed to accompany the textbook *Electricity, Electronics, and Control Systems for HVAC*, 4th edition. The 21 chapters in this lab manual match each of the chapters in the textbook. This lab manual provides additional information to help students master the information and skills they will need when they are working on the job as air-conditioning and refrigeration technicians. Each chapter includes objectives, an introduction, key terms, safety for the exercise, a list of tools and materials for the lab exercise, references to the textbook, and the sequence to complete the lab exercise. Each chapter will have evaluations, which include a matching test to see if students understand the terms for the chapter, a short true or false quiz, and multiple choice questions.

OBJECTIVES

Objectives for each chapter will be provided so that students will know what they are expected to learn in the chapter. The objectives will relate to what the students will learn in the lab exercises.

INTRODUCTION AND OVERVIEW

The introduction will provide an overview of the chapter and a summary of all the material covered in the lab exercise. The overview will provide sufficient information to remind the students what is important in the chapter and how they apply to the lab exercises and the terms for the chapter. The overview is not designed to stand alone; rather, it will remind the students to return to the chapter in the textbook to get more in-depth information where they will find additional pictures and diagrams about the material.

TERMS AND MATCHING TEST

The terms for the chapter will be provided in each chapter of the lab manual and the students will be directed back to the chapter in the textbook to find more information about them. The students will also be encouraged to use the glossary at the end of the textbook to find the definition of the term. A matching test will be provided in each chapter to strengthen the students' vocabulary and understanding of these terms.

QUIZZES

A true or false quiz and multiple choice questions are provided at the beginning of each chapter in this lab manual, and they will provide a way to determine how well the students understand the material. These quizzes will also be available as test bank questions so that instructors can get information back quickly to the students about their scores on the quiz. The quizzes, along with questions that are embedded in the exercises, can be used as post tests for the chapter to ensure that students have an adequate understanding of the material.

LAB EXERCISES

One or more lab exercises will be provided in each chapter to help students learn the fundamental skills they will need to succeed on the job. These skills are identified to prepare the students for tasks they must do to install, troubleshoot, and repair HVAC and refrigeration systems and their component parts. Each lab exercise will have a number of sequential steps that will lead the students through processes that they will be able to repeat when they are on the job. Each exercise will have the following:

Safety for this Lab Exercise
Tools and Materials Needed to Complete the Lab Exercise
References to the Text
Sequence to Complete the Lab Tasks
Checking Out

Safety for this Lab Exercise

Safety for each lab exercise is provided, and will help students to learn to work safely, which is the most important part of any job. The safety instructions will enable students to complete the lab exercise safely as well as help develop safe working habits.

Tools and Materials Needed to Complete the Lab Exercise

This section will provide a complete list of parts and tools that are needed to complete the lab. This will help instructors prepare the lab exercises so that all the material and tools will be available when students start the lab exercises. This will also help instructors determine the amount of consumable material and parts that are needed in addition to the tools that students will need to successfully complete each exercise.

References to the Text

This section will remind students to refer to the chapter in the textbook that provides additional information, diagrams, and pictures. This should help them to better understand the concepts involved in learning the skills to become an HVAC and refrigeration technician.

Sequence to Complete the Lab Tasks

This is a detailed set of step-by-step instructions on how to complete the lab exercise. The students can also use these step-by-step procedures when they get to the job and need to install or troubleshoot parts and systems. When students complete the lab exercise, it will reinforce the knowledge they have learned from reading the textbook and what they have learned in the classroom. Included in this section are questions that are connected to each step of the lab exercise to check understanding of what is being observed. They will help reinforce the activities and allow the instructor to determine to what extent the students understand the information.

Checking Out

One of the most important steps in the lab process is to ensure clean up has been taken care of and that students put all materials, tools, and supplies back to their proper storage locations. Another step that is important is that students need to check with their instructors to ensure they have satisfactorily completed the entire lab, cleaned their area, and it is okay to leave the lab. This section will be provided as the last step of every exercise and the instructor will need to sign off indicating the student has completed the material satisfactorily.

CHAPTER 1

Safety Practices for HVAC

OBJECTIVES

At the end of this chapter you will be able to:

1. Evaluate your work area and develop a safety plan for electrical systems.
2. Evaluate your work area and develop a safety plan for using electrical hand and power tools.
3. Evaluate your work area and develop a safety plan for working around refrigeration systems where you may be exposed to belts, pulleys, and fan blades.
4. Evaluate your work area and develop a safety plan for fire safety.

INTRODUCTION AND OVERVIEW

This chapter explains the safety issues you must be aware of when you are working on the job. This section in the lab manual will provide some activities that will help you observe your own lab area for safety concerns and learn to create a plan to keep you aware of the safety issues and work as safely as possible. You will also learn to use the same process to observe the job sites when you go to work on a residential or commercial air-conditioning or refrigeration system. You will learn in this section that you are responsible for your safety and the safety of co-workers. You will learn about working safely with electrical circuits, using hand and power tools safely, and working safely around mechanical systems such as fans, belts, and pulleys. The material in Chapter 1 of the textbook will help supplement the material and information in this lab manual.

Safety Gear

It is important to always wear safety glasses when you are on the job or in the lab. The safety glasses will protect your eyes against flying debris and from molten metal if an electrical short occurs. When an electrical short circuit occurs, a large amount of current is drawn, which causes any metal contacts or copper wire to turn into a hot molten metal that will travel a distance of over 12 inches and would easily get into your eyes if you are not wearing safety glasses with side shields. If you are working with electric drills or saws, it is possible for material around the bits and blades to be thrown into the air where they can get into your eye. Many injuries result every year from fine slivers of metal getting into a worker's eye, because they were not wearing safety glasses.

You may also be requested to wear a hard hat if you are working on a commercial or residential construction site, and you may need to wear a safety harness if you are working above ground level when you install duct work or are installing a chimney for a furnace that must go through the roof. Be sure to check with the local codes to identify all the safety gear you are required to have in your possession and must wear any time you are on the job.

Electrical Shock Hazards

At times you will have to work on or around live electrical circuits, so you must be aware of the electrical shock hazards. An electrical shock can occur when your skin comes into contact with electrical current and your body becomes a path to ground. When this occurs a small amount of electrical current can flow from the point where you are touching the electrical circuit to the point where your body is grounded. The important point is to limit your exposure to the parts of the electrical circuit where you can receive a shock and increase the amount of resistance between your body and ground. You can limit your exposure to electrical shock hazards by wearing the proper clothing such as gloves or long sleeve shirts and pants to limit the exposure of bare skin to electrical terminals, and you should wear leather boots with rubber soles that increase your electrical resistance to limit your body's ability to carry current to ground. You can refer to Chapter 1 in the textbook to review the amount of current your body can withstand before severe damage is caused or before you receive a fatal electrical shock that can stop your heart.

Administering Cardiopulmonary Resuscitation (CPR)

If a person has received a severe electrical shock, they may stop breathing and their heart may stop beating. Their muscles may also contract and they may not be able to break contact from the point where they are receiving the electrical current. If you are trying to provide aid to a person who has received an electrical shock, you should be sure that you do not touch the victim until you can determine they are no longer in contact with the electrical source or that power has been turned off. The person who has received the electrical shock may have also received severe burns. If the person has stopped breathing and does not have a pulse, any person qualified to give CPR will need to start it as soon as possible and you should call 911 for emergency help as quickly as possible. If you are not qualified to give CPR, you should contact your local provider and take the short course to become qualified.

Electrical Safety Gear and Grounded Equipment

One way to limit the possibility of receiving an electrical shock is to ensure the HVAC equipment you are working on is grounded. You can determine if the equipment is grounded properly by using your voltmeter to test for voltage from an L1 source to the metal cabinet of the equipment. If the equipment is properly grounded, you will measure 110 volts between any L1 source and the metal cabinet. If you do not measure 110 volts between the L1 source and the cabinet, it indicates the system is not properly grounded, and you will need to turn off all electrical power to the system and connect the ground wire to the cabinet and to the system ground at the electrical disconnect.

When the electrical system is properly grounded, any time an electrical power source comes into contact with the metal cabinet it will cause the circuit breaker to trip or the fuse to blow. If the system is not grounded, you can receive a severe electrical shock if you come in contact with the equipment's metal cabinet and your feet are grounded, if you are standing in a wet location, of if the grass around the HVAC equipment is damp or wet from rain or early morning dew. If you are using electrical power tools, you should be sure that you are using a grounded power cord that has the third grounding terminal, plastic case tools, or battery powered tools that do not need to be grounded.

Ground Fault Interrupter (GFI) Circuit Breaker

Another way to limit the possibility of receiving an electrical shock from faulty short circuit is to use ground fault interrupter (GFI) circuit breaker, or use electrical power cords that have a GFI breaker built in. The GFI breaker is designed to measure the amount of current that starts out from the power source and compare it to the amount of current that returns after moving through the circuit. If the amount of current is not exactly the same, it causes the GFI circuit breaker to trip because the difference in current indicates a short circuit exists and some of the current is leaking out of the circuit to a point where it should not be. This is usually the point where a human is receiving an electrical shock, so the GFI breaker interrupts the power source and turns off the power. The GFI has a specific current setpoint, and it is less than the level that can cause injury to a human. The GFI senses and trips when the difference in the current is exceeded. Most electrical codes require a GFI circuit breaker be used to protect any electrical receptacles that are available for outdoor use. You should make sure that any electrical extension cord that you use for power tools has a GFI circuit breaker on it and test it according to the recommendations of the manufacturer.

Fire Safety and Fire Extinguishers

If you work in a shop, on a construction site, or from a truck, you should have a fire extinguisher available at all times. Fire extinguishers are rated by classification. Class A fire extinguishers can be used on ordinary combustible material such as paper, wood, or clothing. This type of fire can easily be extinguished using water. Class B fire extinguishers are designed for fires in flammable liquids, grease, and other materials that can be extinguished by smothering or removing air (oxygen) from the chain reaction. Class C fire extinguishers are used on fires that involve live electrical equipment. The extinguishers for these types of fires must use material that is non-conducting, so that the person using the extinguisher is not exposed to additional hazard from electrical shock when fighting the fire. Class D fire extinguishers are used on fires in combustible metals such as sodium, magnesium, or lithium. Special extinguishers must be used to lower the temperature and remove oxygen from these fires.

Color codes for these fire extinguishers have been developed, and it is important that you recognize the type of fire and use the proper extinguisher. Class A fire extinguishers are green, Class B fire extinguishers are red, Class C fire extinguishers are blue, and Class D fire extinguishers are yellow. It is also important that you recognize the type of material in your shop area and have a specific plan for using fire extinguishers for small fires, and escape plans for evacuation if the fire is too large to contain. Fire extinguishers need to be inspected and tested according to the schedule listed on their label.

The *National Electrical Code®* (NEC®)

The *National Electrical Code®* (NEC®) is written by the National Fire Protection Agency who ensures wiring that you are installing or troubleshooting will not cause a fire or pose a shock risk to those working on it. The NEC® consists of nine chapters. Each chapter has a number of articles that spell out such things as wiring and protection, methods and materials, equipment for general use, special occupancies, special equipment, special conditions, and communication systems. The NEC® also has a number of tables that spell out other conditions. You will learn more about the NEC® in later chapters, and you will be provided with specific examples and tables that you will be expected to look up and use.

NFPA 70E® Compliance: Protecting Technicians against Electrical Arc Flash and Arc Blast

The National Fire Protection Association has developed a comprehensive standard called NFPA 70E® that establishes the best electrical safety practices on how to protect technicians who work with electrical circuits above 50 volts. This standard explains how to protect technicians who work with electrical circuits against electrical arc flash and arc blast exposure. The Occupational

Safety and Health Administration (OSHA) has adopted regulations on safe electrical practices based on NFPA 70E®. The goal of NFPA 70E® is to protect technicians from the hazards of electrical shock, electrocution, arc flash, and arc blast. You can find more information on the Internet about this important safety issue.

Lockout/Tagout Procedures

If you are working on a residential or commercial HVAC or refrigeration system you may need to turn off all power and lock it out with a padlock to ensure the electrical power stays in the off position. The lockout and tagout procedure provides a means of locking out a disconnect switch and putting on a tag with your picture to indicate you are the technician that has locked out the system. When you have completed your repairs, you can remove the padlock and restore power to the system. Having the power turned off and tagging it out ensures the system is in the safest condition for you to work on the electrical system.

Driving Safety

If you are working on the job and running service calls with a service van, you must be aware of safety issues that are specifically related to driving. For example, you should never drive the company vehicle over the speed limit and you should obey all driving laws. You should also be aware when you are parking your vehicle at a residence that small children may play around your truck or leave toys behind it. For this reason it is very important that you walk around the rear of the vehicle prior to starting it and backing it out of the driveway.

TERMS

Cardiopulmonary resuscitation (CPR)
Circuit breaker
Electrical shock
Electromotive force
Fuse
Ground
Ground fault interrupter (GFI)
Grounding adapter
Lockout/tagout
National Electrical Code® (NEC®)
Short circuit

MATCHING

Place the letter A–K for the definition from the list that matches with the terms that are numbered 1–11.

Score __________

1. ____ Cardiopulmonary resuscitation (CPR)
2. ____ Circuit breaker
3. ____ Electrical shock
4. ____ Electromotive force
5. ____ Fuse
6. ____ Ground
7. ____ Ground fault interrupter (GFI)
8. ____ Grounding adapter
9. ____ Lockout/tagout
10. ____ *National Electrical Code*® (NEC®)
11. ____ Short circuit

A. Another term for voltage or potential difference.
B. A set of regulations governing the construction and installation of electrical wiring and apparatus.
C. An electrical device that is specifically designed to protect a circuit against overcurrent and can be reset over and over.
D. A protection device similar to a circuit breaker. It senses small amounts of ground currents and opens the circuit.
E. A special plug adapter that allows a three-prong male plug to plug into a two-prong receptacle.
F. A circuit where resistance in the circuit is near zero and causes extremely high current if voltage is applied.
G. A first-aid technique used to keep a person whose heart has stopped alive while waiting for a more qualified emergency responder to arrive.
H. A safety process where power is turned off and locked out with a padlock.
I. A device designed as a one-time protection against overcurrent or short-circuit current.
J. The earth or any point in an electrical circuit that has the same potential as earth.
K. A condition that occurs when electrical current comes into contact with humans.

TRUE OR FALSE

Place a *T* or *F* in the blank to indicate if the statement is true or false.

Score __________

1. ____ Class I fires include petroleum and solvent-type fires.
2. ____ You can use a Class A fire extinguisher on an electrical fire.
3. ____ A circuit breaker protects the HVAC system from drawing excessive current and damaging its wiring.
4. ____ A GFI (ground fault interrupter) circuit breaker should be used with an extension cord in a damp or wet location.
5. ____ If you are working at an older residence that does not have grounded receptacle outlets, it is safe to use a three-prong extension cord if you break of the ground terminal.

MULTIPLE CHOICE

Circle the letter that represents the correct answer to each question.

Score __________

1. The color of a Class A-type fire extinguisher is ______________ and is used for paper or wood fires.
   a. red
   b. green
   c. yellow

2. A GFI (ground fault interrupter) receptacle is a:
   a. special receptacle that has a third terminal for a ground wire.
   b. special receptacle for power tools that allows more than one power tool to be plugged into it at the same time.
   c. receptacle that has a special circuit breaker built into it that can determine if a short circuit occurs and automatically opens to protect the circuit.

3. If a circuit breaker that protects an air-conditioning system continually trips at a residence, you should:
   a. put in a larger one.
   b. troubleshoot the system to determine what is causing the system to draw extra current.
   c. put a jumper wire around the circuit breaker so it is no longer in the circuit.

4. A three-prong plug:
   a. is better than a two-prong plug because the extra prong helps keep the plug in the socket better.
   b. is not needed if you have double insulated tools.
   c. provides a third prong that is the grounding wire for the system, and ensures that the ground wire is provided to all power tools that are plugged into it.
   d. Both b and c

5. If an electrical power tool has a power cord with a three-prong plug and the electrical outlet has two prongs, you should:
   a. cut the third prong (ground lug) off the power cord so that you can use the outlet.
   b. find an adapter that allows you to plug a three-prong plug into a two-prong outlet.
   c. not use this outlet because the power tool needs to be grounded for safe operation and the two-prong outlet is not grounded. Locate a GFI outlet or a grounded receptacle.

LAB EXERCISE: BEING AWARE OF SAFETY ISSUES ON THE JOB

Introduction

This lab exercise will help you understand the safety problems that you must be aware of when working on HVAC and refrigeration systems. Chapter 1 in the textbook will help you understand electrical safety and other safety issues that you will encounter on the job. You can refer to the material in Chapter 1 in the textbook to help you review this material.

Safety for this Lab Exercise

This lab exercise does not have any activity that poses a safety hazard. You will only be writing a short list of items that refer to safety issues in the workplace and how you can develop a safety plan to work with when you are on the job.

Tools and Materials Needed to Complete the Lab Exercise

You will only need paper and pencil for this lab exercise and you will need to visit your lab area.

References to the Text

Refer to Chapter 1 in the textbook for additional information. You may need to read sections of the chapter again to help you understand the material in this exercise.

Sequence to Complete the Lab Tasks

1. Make a fire safety plan for your lab area and classroom that includes an evacuation plan.

   a. Make a list of all the possible fire hazards for your classroom and lab.

   ____

   ____

   b. Identify an evacuation plan in case of fire. Include where to go, what routes to take, and how to account for everyone when you reach an evacuation point.

   ____

   ____

   c. Identify the types of fire extinguishers in your classroom or lab, and indicate what types of fires they are designed to work on and where they are located.

   ____

   ____

2. Locate and inspect the electrical power tools in your shop. Note any problems with the condition of the power cords and check if the grounding terminal is in good shape. Report on the tools and the condition you have found each in.

   ____

   ____

3. Identify potential electrical shock hazards in your lab and identify where the emergency electrical shut off is located.

   a. Have your instructor walk around your lab and identify any point where you have exposed electrical terminals and list them. Also indicate how you will limit your exposure to electrical shock hazards in your shop.

   ____

   ____

   b. Most HVAC labs have one or more emergency stop push buttons that will shut off all electrical power when they are activated. Have your instructor walk around your lab and identify the location of all emergency electrical shut off buttons. List the locations of all emergency shut off buttons.

   ____

   ____

4. Have your instructor take you on a tour of your lab and identify any equipment or fans that have belts and pulleys or other moving parts that may allow you to become entangled in. List them in the space below.

   ____

   ____

5. Identify any equipment that should be locked out and tagged out in your lab. Identify where the padlocks and tags for the lockout and tagout process are located. Indicate when you should use the lockout and tagout process in your lab or on the job.

____

____

Checking Out

When you have completed this lab exercise, clean up your area, return all tools and supplies to their proper place, and check out with your instructor. Your instructor will initial here to indicate you are ready to check out. _______

CHAPTER 2

Fundamentals of Electricity

OBJECTIVES

At the end of this lab exercise you will be able to:

1. Understand the relationship between volts, ohms, and amperes, and be able to calculate any one of these if a value for the other two is given.
2. Understand the relationship between watts, volts, ohms, and amperes, and be able to calculate any one of these if two of the other values are given.

INTRODUCTION AND OVERVIEW

In this lab exercise you will be provided information about using and understanding Ohm's law. Ohm's law allows you to calculate volts, ohms, or amperes when the value of any two of the three are provided. By calculating the volts, ohms, and amps you will be able to predict what will happen when one of the values stays constant and the other values change. These calculation skills will help you predict what will happen in circuits when resistance increases or decreases as loads are added in series or parallel. These calculation skills will also help you make decisions when you are troubleshooting.

TERMS

Alternating current (AC)
Ammeter
Atom
Clamp-on ammeter
Conductor
Current
Direct current (DC)
Electrical control
Electrical load
Electrical potential
Electrical power
Electricity
Electromotive force (EMF)
Electron
Insulator
Neutron
Nucleus
Ohmmeter
Ohm's law
Proton
Resistance
Static electricity
Voltage
Watts

MATCHING

Place the letter A–V for the definition from the list that matches with the terms that are numbered 1–22.

Score __________

1. ____ Alternating current (AC)
2. ____ Ammeter
3. ____ Atom
4. ____ Clamp-on ammeter
5. ____ Conductor
6. ____ Current
7. ____ Direct current (DC)
8. ____ Electrical control
9. ____ Electrical load
10. ____ Electrical power
11. ____ Electromotive force (EMF)
12. ____ Electron
13. ____ Insulator
14. ____ Neutron
15. ____ Nucleus
16. ____ Ohmmeter
17. ____ Ohm's law
18. ____ Proton
19. ____ Resistance
20. ____ Static electricity
21. ____ Voltage
22. ____ Watts

A. A wire that is usually made of copper or aluminum that carries electrical current.
B. The energy in an electrical circuit that is the result of voltage and current, and the units are watts.
C. The switches in a schematic or wiring diagram that provide the sequence that turn loads on or off.
D. Current that changes from a positive level to a negative level periodically. Its waveform is a sine wave.
E. The devices in an electrical system that consume energy and convert it to motion, heat, or light.
F. An instrument that measures electrical current in amperes.
G. An electrical instrument that measures current and has a set of "jaws" that open to allow them to wrap around a wire.
H. Current that flows in only one direction.
I. A unit of matter. The smallest unit of an element that consists of a nucleus that has a positive charged proton and neutral charged neutron.
J. The flow of electrons that is measured in amperes.
K. The electromotive force or pressure in an electrical circuit that causes electrons (current) to flow.
L. A set of mathematical calculations that show the relationship between voltage, resistance, and current.
M. The units of electrical power.
N. The neutral part of the atom. It is located with the proton as part of the nucleus of the atom.
O. The electrical charge that is caused by the imbalance of positive and negative charges.
P. A material that does not conduct electrical current easily. These materials have very high resistance.
Q. The positive part of an atom.

R. The negative part of an atom that is located in orbits (shells) and move around the nucleus (center) of the atom.
S. The opposition to current flow.
T. Another term for voltage or potential difference.
U. A meter that is designed to measure resistance and indicate the value in ohms.
V. The center of an atom that consists of protons and neutrons.

TRUE OR FALSE

Place a *T* or *F* in the blank to indicate if the statement is true or false.

Score __________

1. ____ The insulation coating on wire has high resistance.
2. ____ Voltage is the flow of electrons.
3. ____ Resistance is the opposition to current flow.
4. ____ Watts are the units for current.
5. ____ A good conductor has low resistance.
6. ____ The negative part of an atom is an electron.
7. ____ Milli is the prefix for 1/1000.
8. ____ A battery creates voltage from a chemical reaction.
9. ____ Voltage is pressure that causes current to flow.
10. ____ Kilo is the prefix for one thousand (1,000).

MULTIPLE CHOICE

Circle the letter that represents the correct answer to each question.

Score __________

1. A valence electron is:
   a. the electron closest to the nucleus.
   b. the electron in the outermost shell.
   c. an electron with a positive charge.

2. Wattage is the unit for:
   a. power.
   b. current.
   c. voltage.

3. Resistance is:
   a. the force that moves electrons.
   b. wattage.
   c. the opposition to current flow.

4. Current will _______________ when voltage increases and resistance stays the same.
   a. increase
   b. decrease
   c. stay the same

5. Wattage will _______________ when voltage increases and current stays the same.
   a. increase
   b. decrease
   c. stay the same

6. A milliamp is:
   a. one-thousandth of an ampere.
   b. one-millionth of an ampere.
   c. one million amps.

7. Like charges:
   a. attract.
   b. repel.
   c. have no effect on each other.

8. The atom has which of the following parts?
   a. electron
   b. proton
   c. neutron
   d. nucleus
   e. All the above

9. A conductor:
   a. is a material that stops the flow of electrons.
   b. is a material that allows electrons to flow easily.
   c. provides a source of voltage for a circuit.

10. A load is:
    a. the part of a circuit that has resistance.
    b. a switch in a circuit that controls current flow.
    c. a power source for a circuit.

LAB EXERCISE: UNDERSTANDING OHM'S LAW

Safety for this Lab Exercise

The activities for this lab use only paper and pencil, so there are no safety concerns for this lab exercise.

Tools and Materials Needed to Complete the Lab Exercise

The only materials needed for this lab exercise are paper, pencil, and calculator.

References to the Text

Refer to Chapter 2 in the textbook for additional information. You may need to read sections of the chapter again to help you understand the material in this exercise.

Sequence to Complete the Lab Task

Ohm's Law

1. Referring to the Ohm's law pie, identify and write a statement for each letter indicating what it means. Remember the unit for voltage is volts, the unit for current is amperes, and the unit for resistance is ohms. When you are calculating voltage, current, and resistance you should use the letter *E* for electromotive force (volts), *I* for intensity (current), and *R* for resistance (ohms). When you have calculated a number and have an answer to a problem, you should use the units voltage, amperes, and ohms instead of the letters *E*, *I*, and *R*.

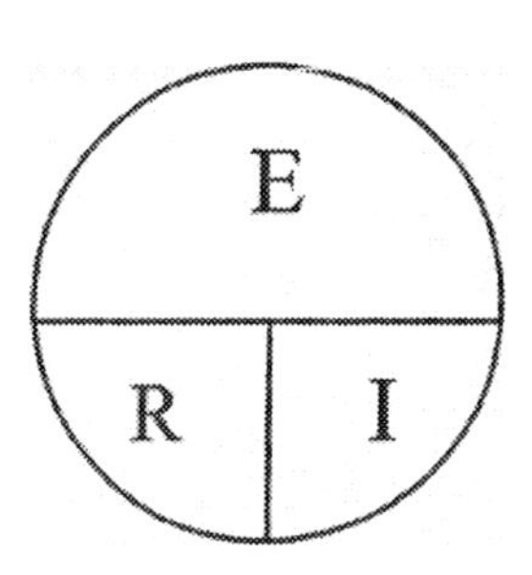

R = ____________________

E = ____________________

I = ____________________

2. Use the Ohm's law pie and identify the three formulas. Write the formula for volts, amperes, and ohms, and place each beside the letter provided below.

Formula # 1	*Formula # 2*	*Formula # 3*
E = ________	I = ________	R = ________

3. Use Ohm's law to calculate the following.
   a. Find voltage if current is 6 amps and resistance is 10 ohms.

      ________ Voltage

   b. Find current if voltage is 12 volts and resistance is 6 ohms.

      ________ Amps

   c. Find resistance if voltage is 24 volts and current is 2 amps.

      ________ Ohms

4. Use Ohm's law to calculate the following, and then answer the follow-up question in d.
   a. Find current if voltage is 12 volts and resistance is 2 ohms.

      ________ Amps

b. Find current if voltage is 12 volts and resistance is 3 ohms.

__________ Amps

c. Find current if voltage is 12 volts and resistance is 4 ohms.

__________ Amps

d. What can you predict will happen to current if voltage is held constant and resistance is increased?

____

____

5. Use Ohm's law to calculate the following, and then answer the follow-up question in d.

a. Find voltage if current is 2 amps and resistance is 2 ohms.

__________ Volts

b. Find voltage if current is 2 amps and resistance is 4 ohms.

__________ Volts

c. Find voltage if current is 2 amps and resistance is 6 ohms.

__________ Volts

d. What can you predict will happen to voltage if current is held constant and resistance is increased?

____

____

Sequence to Complete the Lab Task

Watt's Law

1. Referring to the Watt's law pie, identify and write a statement for each letter indicating what it means. Remember the unit for power is watts, the unit for voltage is volts, and the unit for current is amperes. When you are calculating power, voltage, and current, you should use the letter *E* for electromotive force (volts), *I* for intensity (current), and *P* for power (watts). When you have calculated a number and have an answer to a problem, you should use the units watts, voltage, and amperes instead of the letters *P*, *E*, and *I*.

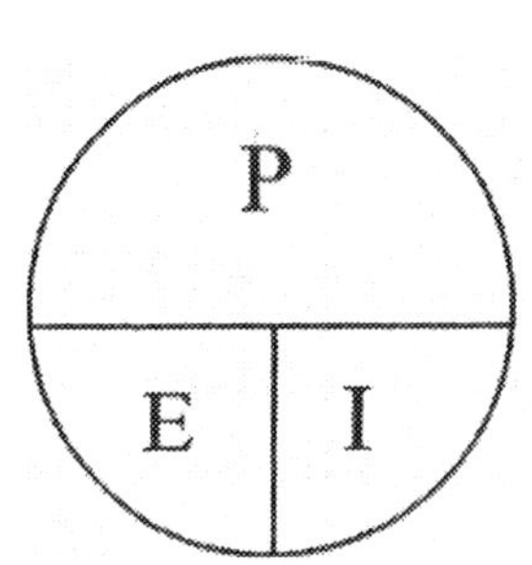

P = ____

E = ____

I = ____

2. Use the Watt's law pie and identify the three formulas. Write the formula for volts, amperes, and ohms, and place each beside the letter provided below.

Formula # 1	*Formula # 2*	*Formula # 3*
P = ________	E = ________	I = ________

3. Use Watt's law to calculate the following.

   a. Find wattage if current is 6 amps and voltage is 10 volts.

   ________ Watts

   b. Find current if voltage is 12 volts and wattage is 36 watts.

   ________ Amps

   c. Find voltage if wattage is 24 watts and current is 2 amps.

   ________ Volts

4. Use Watt's law to calculate the following, and then answer the follow-up question in d.

   a. Find current if wattage is 60 watts and voltage is 10 volts.

   ________ Amps

   b. Find current if wattage is 60 watts and voltage is 20 volts.

   ________ Amps

   c. Find current if wattage is 60 watts and voltage is 30 volts.

   ________ Amps

   d. What can you predict will happen to current if wattage is held constant and voltage is increased?

   ____

   ____

5. Use Watt's law to calculate the following, and then answer the follow-up question in d.

   a. Find wattage if voltage is 60 volts and current is 2 amps.

   ________ Watts

   b. Find wattage if voltage is 60 volts and current is 10 amps.

   ________ Watts

   c. Find wattage if voltage is 60 volts and current is 20 amps.

   ________ Watts

   d. What can you predict will happen to wattage if voltage is held constant and current is increased?

   ____

   ____

Checking Out

When you have completed this lab exercise, clean up your area, return all tools and supplies to their proper place, and check out with your instructor. Your instructor will initial here to indicate you are ready to check out. _______

Chapter 3

Electrical Circuits

OBJECTIVES

At the end of this lab exercise you will be able to:

1. Explain what happens to current when a switch that is wired in a series is opened.
2. Identify components that are wired in series in an HVAC circuit.
3. Identify components that are wired in parallel in an HVAC circuit.
4. Explain what happens if an open occurs anywhere in a series circuit.
5. Explain why loads are wired in parallel instead of series.

INTRODUCTION AND OVERVIEW

In this lab exercise you will be provided information about series circuits and parallel circuits. You will learn a number of things that will help you troubleshoot HVAC circuits, and predict what will happen in an HVAC circuit if a wire, switch, or component has an open. You will also learn what happens to current in a series-parallel circuit, which will help you understand how fuses and wires are sized. You will find that understanding how components operate when they are wired in series, and how they operate when components are wired in parallel, will help you understand troubleshooting much better. Most HVAC circuits have some switchers and controls wired in series and they will have loads wired in parallel.

TERMS

Closed circuit
Closed switch
Control circuit
Good fuse
Open circuit
Open fuse
Open switch
Parallel circuit
Power circuit
Series circuit
Series-parallel circuit
Voltage drop

MATCHING

Place the letter A–L for the definition from the list that matches with the terms that are numbered 1–12.

Score __________

1. ____ Electrical circuit
2. ____ Closed switch
3. ____ Control circuit
4. ____ Good fuse
5. ____ Open circuit
6. ____ Open fuse
7. ____ Open switch
8. ____ Parallel circuit
9. ____ Power circuit
10. ____ Series circuit
11. ____ Series-parallel circuit
12. ____ Voltage drop

A. The part of an HVAC circuit that has the evaporator fan motor, condenser fan motor, and compressor motor connected to it.
B. The part of an electrical diagram that contains the switches that control when the loads (motors or electrical resistance coils) turn on or off.
C. A circuit that has a point in it that has extremely high resistance that stops the flow of electrical current.
D. A switch that has high resistance between its contacts.
E. A circuit that has a power source, a number of conductors, and at least one load.
F. A fuse that allows current to flow through it and has very low resistance.
G. A circuit that has two or more loads connected so that each load receives the same voltage.
H. A circuit that has components connected both in series and in parallel.
I. A switch that has its contacts closed and the resistance between the contacts is low, which allows current flow through it.
J. A fuse that has extremely high resistance and will not pass current.
K. The voltage that occurs when current flows through a resistor or load in an electrical circuit.
L. A circuit that has only one path. Any time there is an open anywhere in the circuit, all current flow stops.

TRUE OR FALSE

Place a *T* or *F* in the blank to indicate if the statement is true or false.

Score __________

1. ____ When a series circuit has an open, current flow stops in all parts of the circuit.
2. ____ The main reason the switches are connected in series is so that if any one of the switches opens, current to the load will be interrupted.
3. ____ It is a good practice to connect the fan motor and compressor motor for an air conditioner in series so that when one quits the other will stop.
4. ____ Two heating elements should be connected in series if you want them to split the amount of voltage applied to them.
5. ____ All resistors have the same amount of resistance if they are the same physical size.
6. ____ Current in each part of a parallel circuit will always be the same.
7. ____ Voltage in the branches of parallel circuits will always be the same.
8. ____ Total resistance in a parallel circuit becomes smaller as more resistors are added in parallel.
9. ____ Total current in a parallel circuit becomes smaller as more resistors are added in parallel.
10. ____ The prefix *milli-* (m) means one-millionth.

MULTIPLE CHOICE

Circle the letter that represents the correct answer to each question.

Score __________

1. In a series circuit the current in each part of the circuit is:
   a. always zero.
   b. the same.
   c. equal to the voltage.

2. The amount of resistance in a resistor:
   a. is the same if the resistors are the same size.
   b. can be determined by color codes.
   c. changes as the amount of current that flows through it changes.

3. When three resistors are connected in series, the amount of voltage that is measured across each one is:
   a. determined by the amount of current flowing through it and the amount of resistance it has.
   b. determined by the wattage rating of each resistor.
   c. impossible to determine by calculations.

4. If a thermostat, high-pressure switch, and oil pressure switch are connected in series with a compressor motor, and the oil pressure switch is opened because of low oil pressure, the compressor motor will:
   a. still run because two of the other switches are still closed.
   b. will not be affected because no other loads are connected to it in series.
   c. will stop running because current flow will be zero.

5. If the resistance of a heating element increases from 2 Ω to 5 Ω, the current it draws will:
   a. increase if the voltage stays the same.
   b. decrease if the voltage stays the same.
   c. not change if the voltage does not change.

6. Current in a parallel circuit:
   a. increases as additional resistors are added in parallel.
   b. decreases as additional resistors are added in parallel.
   c. may increase or decrease when resistors are added in parallel depending on their size.

7. Voltage in a parallel circuit:
   a. increases as additional resistors are added in parallel.
   b. decreases as additional resistors are added in parallel.
   c. stays the same across parallel branches as additional resistors are added in parallel.

8. Resistance in a parallel circuit:
   a. increases as additional resistors are added in parallel.
   b. decreases as additional resistors are added in parallel.
   c. may increase or decrease when resistors are added in parallel depending on their size.

9. When one branch circuit of a multiple branch parallel circuit develops an open, voltage in other branch circuits:
   a. decreases to zero.
   b. increases because fewer resistors are using up the voltage.
   c. stays the same as the supply voltage.

10. Electrical meters and circuits primarily use the exponents 10^6, 10^3, 10^{-3}, and 10^{-6} because:
    a. these numbers are easier to use than other exponents.
    b. these exponents are the values for the prefixes M, k, m, and μ.
    c. exponents must be multiples of 3 or 6.

LAB EXERCISE: UNDERSTANDING SERIES AND PARALLEL COMPONENTS AND CIRCUITS

Safety for this Lab Exercise

The activities for this lab uses a number of switches and loads that you will be connecting in series or parallel and applying power. You must be very careful when power is applied to a circuit and avoid coming into contact with any sources of voltage because you can receive an electrical shock. You will be asked to have your instructor check over your circuit before you apply voltage and initial your lab sheet to indicate it is safe to apply power to your circuit. Be sure to turn off all power sources and keep power turned off any time you are making changes to your circuit. You will only turn power on for short periods of time to observe the circuit or to make a meter reading for troubleshooting.

Tools and Materials Needed to Complete the Lab Exercise

You will need a source of voltage for the circuits in this exercise. You will need to match the voltage supply with the components that you are using. You will need several light bulbs and switches for this circuit, which can be AC or DC. If you are using a DC power supply, you will need light bulbs and switches that are rated for DC voltage. If you are using a 120-volt AC power supply, you will need to be sure your switches and light bulbs are rated for 120 volts AC.

1. Three single-pole, single-throw switches.

2. Three light bulbs. (Two of the bulbs should be the same wattage, and one should be a higher or lower wattage.)

3. Wires to connect the components and switches to each other and to the power supply.

4. You will need a board or some method of ensuring that the switches and light bulbs are securely mounted when wires are connected and power is applied so they do not become an electrical shock hazard.

References to the Text

Refer to Chapter 3 in the textbook for additional information. You may need to read sections of the chapter again to help you understand the material in this exercise.

Sequence to Complete the Lab Task

Understanding Switches that Are Connected in Series

1. Figure 3–1 shows the wiring diagram of an electrical circuit with three switches connected in series with one lamp. Figure 3–2 shows the schematic diagram of the same circuit. You can use either diagram to wire this circuit. A series circuit has only one path for the current to flow to reach the lamp. Any time there is an open in the series circuit, current will stop flowing to the lamp. Be sure to keep the power turned off and connect the switches to each other and to the lamp as shown in the diagram. Have your instructor check your circuit to ensure it is connected correctly before you apply power. Your instructor should initial in the space provided when your circuit has been checked to ensure it is wired correctly. _______

2. After you have applied power to your circuit, you should close all three switches and the lamp should become illuminated. If it does not, call your instructor to check your switches, lamp, and wiring. When the lamp is illuminated, you can continue to the next step. *Be aware that there are exposed electrical connections on this circuit and you should not touch any of them as you can receive a severe electrical shock if you come into contact with any of these points.*

3. Now that the lamp is illuminated, open SW2 and report what happens to the lamp.

   ____

   ____

4. Turn switch SW2 back to the on position and open switch SW1. Report what happens to the lamp.

   ____

   ____

5. Turn switch SW1 back to the on position and open switch SW3. Report what happens to the lamp.

   ____

   ____

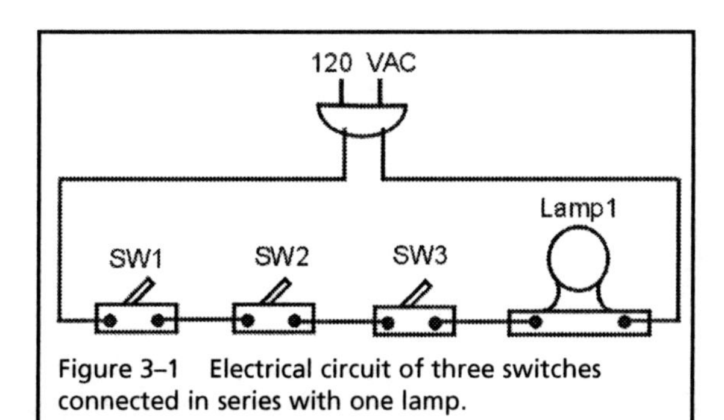

Figure 3–1 Electrical circuit of three switches connected in series with one lamp.

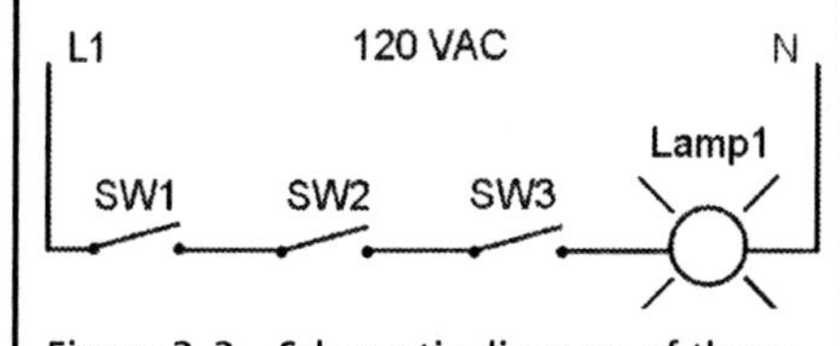

Figure 3–2 Schematic diagram of three switches connected in series with one lamp.

6. Now that you have observed the lamp when you have opened each of the switches, what can you explain about a series circuit when any of the switches are placed to the open position?

____

____

7. Now that you understand the operation of a series circuit, what would you predict would happen to the lamp if all the switches are closed and one of the wires in this circuit developed an open?

____

____

8. Explain how you can use this knowledge of series circuits if you are troubleshooting an HVAC circuit that has a motor and several switches that are connected in series with it and the motor will not turn on.

____

____

Sequence to Complete the Lab Task

Understanding Light Bulbs that Are Connected in Series

1. Figure 3–3 shows the wiring diagram of an electrical circuit with a switch connected in series with two lamps. Figure 3–4 shows the schematic diagram of the same circuit. You can use either diagram to wire this circuit. Typically you will not find two loads such as lamps, relay coils, or motors connected in series, as the voltage for the circuit is split between the two loads rather than each load getting full voltage. If the two lamps have identical wattage, the loads will receive half the voltage they need and they will not operate correctly. If one load has a different wattage than the other, the voltage will split between the loads unevenly. This exercise will show you that the lamps will not work correctly when they are wired in series, and you will see that loads such as motors cannot be wired in series. Be sure to keep the power turned off and connect the switches to each other and to the lamp as shown in the diagram. Have your instructor check your circuit to ensure it is connected correctly before you apply power. Your instructor should initial in the space provided when your circuit has been checked to ensure it is wired correctly.

____

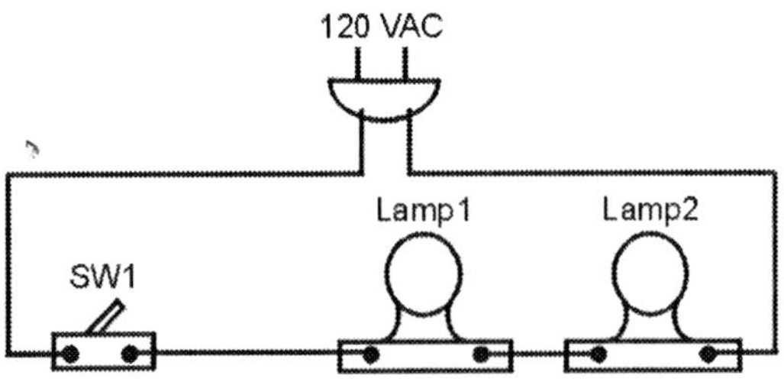

120 VAC
Lamp1
Lamp2
SW1

Figure 3–3 Wiring diagram of two lamps wired in series with each other.

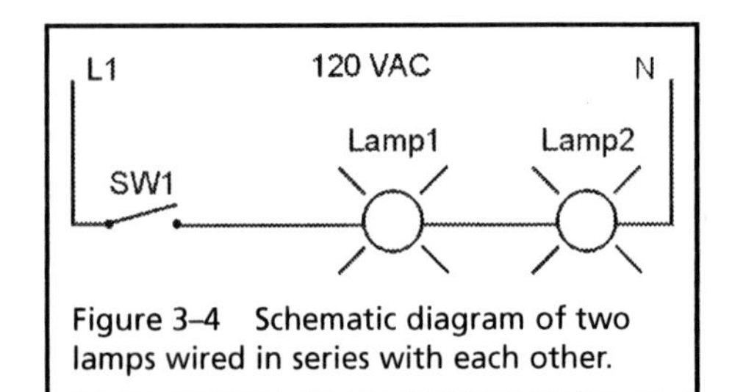

Figure 3–4 Schematic diagram of two lamps wired in series with each other.

2. After you have applied power to your circuit, you should close the switch and both lamps should become illuminated. If the circuit does not operate, call your instructor to check your switch, lamps, and wiring. When the lamps are illuminated, you can continue to the next step. *Be aware that there are exposed electrical connections on this circuit and you should not touch any of them as you can receive a severe electrical shock if you come into contact with any of these points.*

3. Now that the lamps are illuminated, compare how bright each lamp is illuminated as compared to the previous exercise when only one lamp was used.

____

____

4. Turn switch SW1 off and let the lamps cool down. Unscrew one of the two lamps so it will not illuminate and turn SW1 back to the on position. Observe what happens to the lamps.

____

____

5. Turn switch SW1 off and let the lamps cool down. Unscrew one of the two lamps and replace it with a lamp that has different wattage. Turn SW1 back to the on position and observe what happens to the lamps that have different wattage.

____

____

6. Now that you have observed this circuit with both lamps wired in series, explain the problems of wiring two loads such as two motors in series with each other.

____

____

Sequence to Complete the Lab Task

Understanding Switches that Are Connected in Parallel

1. Figure 3–5 shows the wiring diagram of an electrical circuit with three switches connected in parallel with one lamp. Figure 3–6 shows the schematic diagram of the same circuit. You can use either diagram to wire this circuit. A parallel circuit has more than one path for the current to flow to reach the lamp. Current can flow to the lamp when any one of the three switches is closed. Be sure to keep the power turned off and connect the switches to each other and to the lamp as shown in the diagram. Have your instructor check your circuit to ensure it is connected correctly before you apply power. Your instructor should initial in the space provided when your circuit has been checked to ensure it is wired correctly. _______

2. After you have applied power to your circuit, you should close all three switches and the lamp should become illuminated. If it does not, call your instructor to check your switches, lamp, and wiring. When the lamp is illuminated, you can continue to the next step. *Be aware that there are exposed electrical connections on this circuit and you should not touch any of them as you can receive a severe electrical shock if you come into contact with any of these points.*

3. Now that the lamp is illuminated, open all three switches. Report what happens to the lamp.

____

____

4. Turn switch SW2 to the on position and leave SW1 and SW3 open. Report what happens to the lamp.

____

____

5. Turn switch SW1 to the on position and open switches SW2 and SW3. Report what happens to the lamp.

____

____

6. Turn switch SW3 to the on position and open switches SW1 and SW2. Report what happens to the lamp.

____

____

7. Now that you have observed the lamp when you have opened each of the switches, what can you explain about a parallel circuit when any of the switches are placed to the open position?

____

____

8. Now that you understand the operation of a parallel circuit, explain what you must do to get the lamp to turn off.

____

____

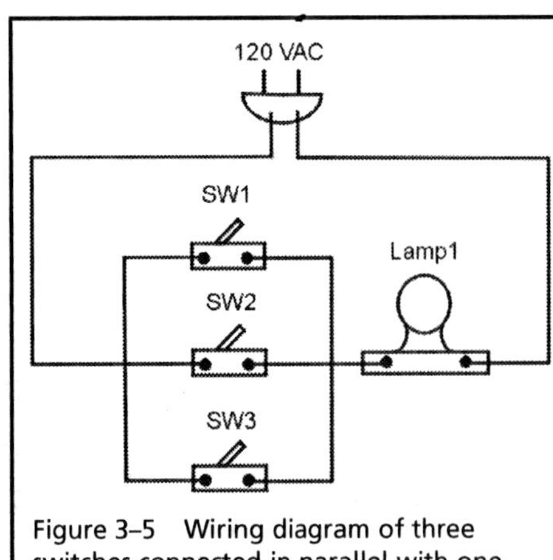

Figure 3–5 Wiring diagram of three switches connected in parallel with one lamp.

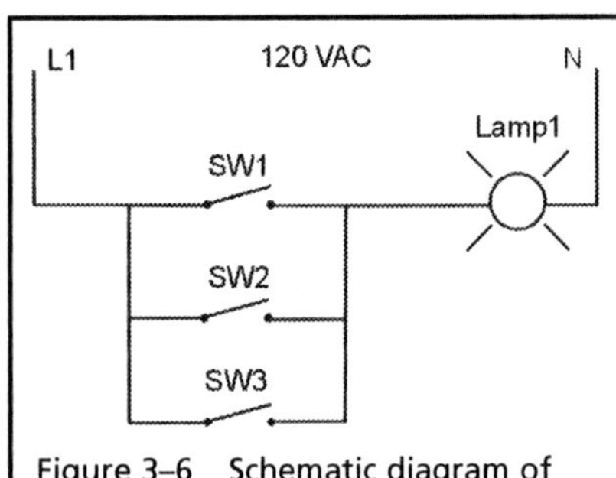

Figure 3–6 Schematic diagram of three switches connected in parallel with one lamp.

9. Explain how you can use this knowledge of parallel circuits if you are troubleshooting an HVAC circuit that has a motor and several switches that are connected in parallel with it and the motor will not turn on.

____

____

Sequence to Complete the Lab Task

Understanding Light Bulbs that Are Connected in Parallel

1. Figure 3–7 shows the wiring diagram of an electrical circuit with one switch connected to two lamps that are connected in parallel. Figure 3–8 shows the schematic diagram of the same circuit. You can use either diagram to wire this circuit. A circuit with loads connected in parallel ensures that the same amount of voltage flows to each lamp. When loads are connected in parallel, each load will receive the same voltage and this ensures each load will operate correctly. Be sure to keep the power turned off and connect the switch to each of the lamps as shown in the diagram. Have your instructor check your circuit to ensure it is connected correctly before you apply power. Your instructor should initial in the space provided when your circuit has been checked to ensure it is wired correctly. _______

2. After you have applied power to your circuit, you should close the switch and both lamps should become illuminated. If they do not, call your instructor to check your switch, lamps, and wiring. When the lamps are illuminated, you can continue to the next step. *Be aware that there are exposed electrical connections on this circuit and you should not touch any of them as you can receive a severe electrical shock if you come into contact with any of these points.*

3. Report what happens to the lamps when they are wired in parallel.

____

____

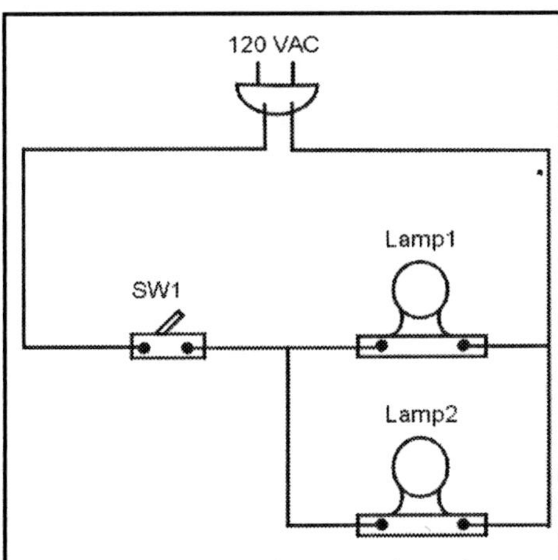

Figure 3–7 Wiring diagram of two lamps connected in parallel and controlled by one switch.

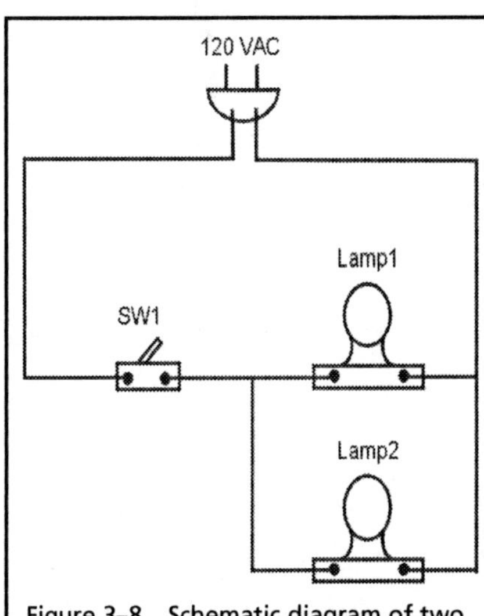

Figure 3–8 Schematic diagram of two lamps connected in parallel and controlled by one switch.

4. Now that the lamps are illuminated, open the switch and report what happens to the lamps.

____

____

5. Now that you have observed the lamps when the switch is opened, what can you explain about the two lamps that are wired in parallel when the switch is placed to the open position?

____

____

6. Explain how you can use this knowledge of parallel circuits if you are troubleshooting an HVAC circuit that has two motors wired in parallel controlled by a single switch, and the motors will not turn on.

____

____

Sequence to Complete the Lab Task

Understanding an HVAC Circuit with Loads and Switches Wired in Both Series and Parallel

Figure 3–9 shows a typical diagram of an HVAC circuit. You should notice that each of the motors and switches are identified with their names and the amount of current they draw. Answer the following questions referring to the diagram.

1. Name the motors that get their current through the cooling thermostat.

____

____

2. Name the motors that get their current through the high-pressure switch and the oil pressure switch.

____

____

3. Is the high-pressure switch and the oil pressure switch wired in series or parallel? __________

4. Is the compressor motor and condenser fan motor wired in series or parallel? __________

5. What will happen to this circuit if the cooling thermostat has a bad contact and cannot pass any current?

____

____

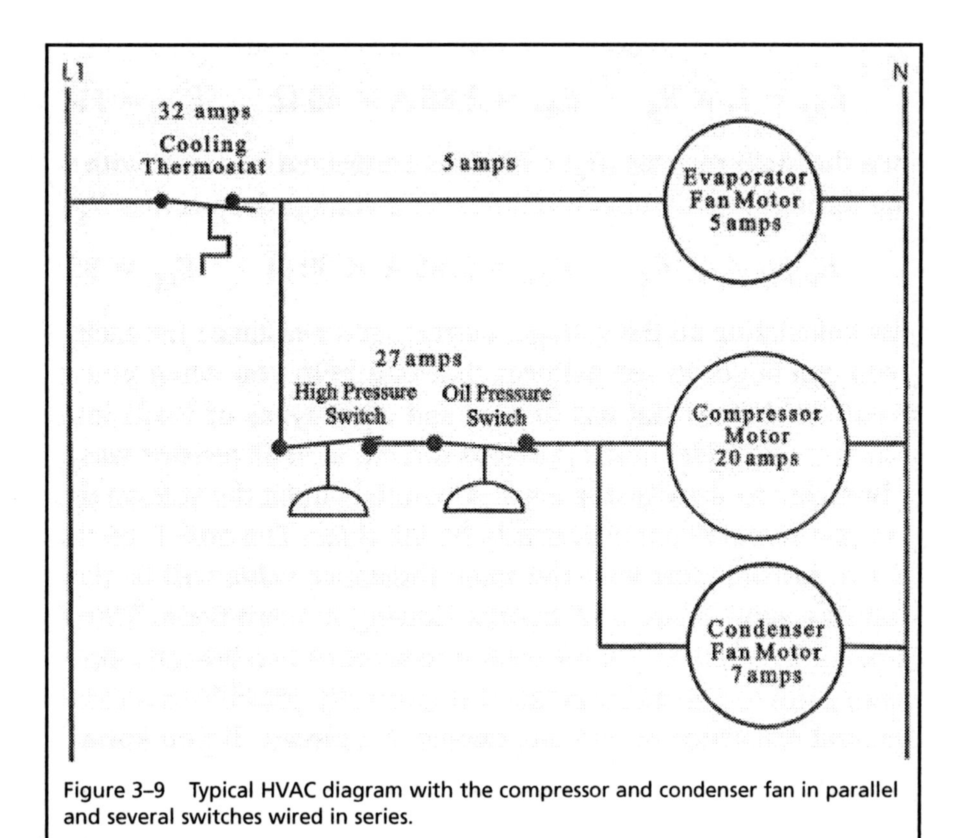

Figure 3–9 Typical HVAC diagram with the compressor and condenser fan in parallel and several switches wired in series.

6. What will happen in this circuit if the system gets low oil pressure and the oil pressure switch opens?

____

____

7. What will happen in this circuit if all the motors are running and the compressor motor develops an open in its winding and stops drawing current?

____

____

Checking Out

When you have completed this lab exercise, clean up your area, return all tools and supplies to their proper place, and check out with your instructor. Your instructor will initial here to indicate you are ready to check out. _______

CHAPTER 4

Meters and Tools for HVAC

OBJECTIVES

At the end of this lab exercise you will be able to:

1. Use an analog voltmeter to measure voltage.
2. Use a digital voltmeter to measure voltage in a variety of components in HVAC circuits.
3. Use a clamp-on ammeter to measure current in a variety of components in HVAC circuits.
4. Use a digital or analog ohmmeter to measure resistance in a variety of HVAC components.

INTRODUCTION AND OVERVIEW

In this lab exercise you will learn the important skills of reading a voltmeter, ammeter, and ohmmeter. Meter reading skills are an important part of troubleshooting for an HVAC and refrigeration service technician. You will need to be able to use both digital and analog meters to measure voltage, current, and resistance in a number of circuits and components. For example, you may need to measure the voltage at a disconnect switch, or across the terminals of motors and other loads in a circuit. You may also need to measure the current that is drawn by various motors in the HVAC system to ensure they are working correctly, and to ensure fuses are sized correctly. Finally, you will be expected to measure resistance in motor windings, transformer windings, and relay coils. You will also be expected to make continuity tests of wires, contacts, and other devices such as fuses. This lab exercise will help you understand the correct way to make these measurements and how to read a meter accurately.

TERMS

Ammeter
Analog meter
Clamp-on ammeter
Digital meter
Magnet
Magnetic field
Meter
Meter movement
Milliammeter
Ohmmeter
Voltmeter

MATCHING

Place the letter A–K for the definition from the list that matches with the terms that are numbered 1–11.

Score __________

1. ____ Ammeter
2. ____ Analog meter
3. ____ Clamp-on ammeter
4. ____ Digital meter
5. ____ Magnet
6. ____ Magnetic field
7. ____ Meter
8. ____ Meter movement
9. ____ Milliameter
10. ____ Ohmmeter
11. ____ Voltmeter

A. Any metal that has its dipoles aligned, or any coil that produces flux lines when current flows through it.
B. A device that can be analog or digital that can measure voltage, current, resistance, or wattage.
C. An instrument that measures electrical current in amperes.
D. An ammeter that has a set of "jaws" that opens to allow them to wrap around a wire. The jaws close around the wire and function as a transformer coil to sense the amount of current flowing through the wire.
E. Field-created flux lines that emanate from the poles of a permanent magnet or from other flux lines of an electromagnet.
F. A meter (usually with a meter movement) that can display the value being measured in a continuous manner by moving a meter needle across a numbered display.
G. An ammeter that measures current down to one-thousandth or 0.001 of an ampere.
H. An electric meter that is able to measure voltage, current, resistance, power, and frequency, and displays the measured values as numbers.
I. A meter that is designed to measure resistance and indicate the value in ohms.
J. An electrical meter that is designed to measure voltage.
K. The internal part of an analog meter that measures current and causes a needle to move across the face of the meter.

TRUE OR FALSE

Place a *T* or *F* in the blank to indicate if the statement is true or false.

Score __________

1. ____ An insulated fuse puller should always be used to remove and replace a fuse in a fused disconnect rather than use metal pliers.
2. ____ All voltage should be turned off when making a resistance measurement.
3. ____ Infinity is the highest resistance reading for an ohmmeter.
4. ____ A VOM meter must be zeroed prior to making voltage, current, or resistance measurements.

5. ____ The voltage should be turned off for safety reasons when making a voltage measurement.

6. ____ The ohmmeter uses an internal battery to supply the power for all resistance readings.

MULTIPLE CHOICE

Circle the letter that represents the correct answer to each question.

Score __________

1. A milliamp is:
   a. one-thousandth of an ampere.
   b. one-millionth of an ampere.
   c. one million amps.

2. When making a current measurement with a VOM meter, you should:
   a. place the meter probes across the load.
   b. turn off all power to the circuit during the current reading so you don't get shocked.
   c. create an open in the circuit and place the meter probes so that the meter is in series with the load.

3. When making a voltage measurement, you should:
   a. place the meter probes across the load.
   b. turn off all power to the circuit during the voltage reading so you don't get shocked.
   c. create an open in the circuit and place the meter probes so that the meter is in series with the load.

4. When making a resistance measurement, you should:
   a. keep all voltage applied to the component you are measuring so you can determine the true resistance.
   b. turn off all power to the circuit and isolate the component you want to measure.
   c. create an open in the circuit and place the meter probes so that the meter is in series with the load.

5. When you are measuring current with a clamp-on ammeter, you should:
   a. open the jaws of the meter and place them around the wire where you want to measure the current.
   b. create an open in the circuit and place the meter probes so that the meter is in series with the load.
   c. turn off all power to the circuit so that you don't get shocked.

6. You would use ______________ to strip a wire to remove its insulation.
   a. wire cutters
   b. common pliers
   c. wire strippers
   d. side-cut pliers

7. Offset screwdrivers are used to:
    a. provide more torque than a regular screwdriver.
    b. tighten or loosen hard-to-reach screws that have small clearance or are in tight spaces.
    c. reach screws in deep recesses.
    d. All of the above

8. A nut driver is a:
    a. type of pliers.
    b. type of wrench.
    c. tool used to strip wires.
    d. special offset screwdriver.

LAB EXERCISE: MEASURING VOLTS, OHMS, AND AMPS

Safety for this Lab Exercise

The exercises for this lab uses a number of circuits and components that are connected to voltage source and have power applied. It is very important that your instructor set up the equipment where you will make these measurements and control the safety of these exercises. You must be very careful when power is applied to a circuit and avoid coming into contact with any sources of voltage because you can receive an electrical shock. You will be asked to have your instructor check over your circuit before you apply voltage, and initial your lab sheet to indicate it is safe to apply power to your circuit when you are ready to make the measurements. Be sure to turn off all power sources and keep power turned off any time you are making changes to your circuit and when you are measuring resistance and making continuity tests. You will only turn power on for short periods of time to observe the circuit or to make a meter reading for troubleshooting.

Tools and Materials Needed to Complete the Lab Exercise

Your instructor will need to set up a number of HVAC systems where you can safely make voltage and current tests when the system is running. It is important for your instructor to set up systems where you can reach the equipment terminals safely with limited exposure to the voltage. You must be fully supervised by your instructor at every step of this lab exercise when power is applied. You can work on your own without supervision when you are measuring the resistance of motors and other devices with coils where no power is applied. You can also make the continuity tests on wires, fuses, and coils without supervision.

You will need the following equipment for this lab exercise:

1. Digital and analog voltmeter
2. Clamp-on ammeter
3. Ohmmeter (can be digital or analog)
4. Several operating HVAC systems that allow easy access to make voltage and current measurements while the system is running
5. Several compressors, open-type motors, transformers, and relay coils that are not connected to power so that you can measure the resistance of their windings
6. Several fuses, switches, and wires that you can test for continuity

References to the Text

Refer to Chapter 4 in the textbook for additional information. You may need to read sections of the chapter again to help you understand the material in this exercise.

Sequence to Complete the Lab Task

Testing Continuity of Fuses, Switch Contacts, and Wires

When you are working on the job and troubleshooting different systems, you will need to be able to make continuity tests of fuses, switches, and wires. *Continuity* is defined as a test of resistance for devices where the amount of resistance being checked is either near zero ohms or very high (infinity). For example, when you are measuring the continuity of a fuse, the resistance will be near zero ohms if the fuse or device is good, and the resistance will be very high (infinity) if the fuse or device is open. Your instructor will provide a number of fuses, switches, and wires for you to measure the continuity. Analog meters have the value zero on the lower end of the scale, and the symbol for infinity (∞), which looks like the number eight on its side. If you are using a digital meter you can check to see what the meter displays when the meter is set on resistance and the meter leads are touching each other, which represents low resistance. Usually, the meter display will show a zero or a very low number. If you set the meter on "audio indication" the meter will also "beep" or make a noise when the resistance being measured is zero or very low. You can determine what the digital meter displays when it is measuring high resistance (infinity). Usually, the meter will show the highest value the meter can read and the value will flash on and off on the display.

You can get the meter to display the value for infinity by setting the meter range to resistance and then holding the two leads so they do not touch. When the meter leads are not touching it represents the condition the meter will have when it is measuring very high (infinite) resistance. Next, you should touch the meter leads together, and the display should show zero or very low resistance, which we interpret as continuity.

SAFETY NOTICE: Be sure that you always ensure that power is turned off to the components that you are testing for resistance of continuity. It is desirable to have the component isolated and not connected in the circuit when you are making resistance or continuity tests.

1. Your instructor will provide a number of components that you can measure the continuity of. Use your meter to measure the continuity of these components and indicate whether the resistance is zero or infinite.

   a. Fuse 1 __________

   b. Fuse 2 __________

   c. Fuse 3 __________

   d. Switch contacts 1 __________

   e. Switch contacts 2 __________

   f. Switch contacts 3 __________

   g. Wire 1 __________

   h. Wire 2 __________

Sequence to Complete the Lab Task

Measuring the Resistance of Motor Windings, Transformer Coils, and Relay Coils

1. The second type of measurement that you will make with an ohmmeter on the job is to measure the exact amount of resistance of motor windings, transformer windings, and relay coils. When you are measuring the resistance of a motor winding, transformer winding, or relay coil, you will need to measure the amount of resistance accurately. In this exercise, you will be provided a number of motors, transformers, and relays for you to measure. These devices are not connected to power and you should ensure that no power is present during these tests. Measure the resistance in the coils of the devices your instructor has provided and write the amount or resistance in the space next to the name of the device. Your instructor will provide answers to the amount of resistance in each winding after you have written your answers in the spaces provided so you can see how accurate your measurements are.

   SAFETY NOTICE: Make sure the power is turned off to all of these components when you are making these tests.

   a. Motor winding 1 (start winding of compressor from terminal S to C) __________

   b. Motor winding 2 (run winding of compressor from terminal R to C) __________

   c. Motor winding 3 (start winding of condenser fan motor from terminal S to C) __________

   d. Motor winding 4 (run winding of condenser fan motor from terminal R to C) __________

   e. Primary winding of transformer 1 __________

   f. Secondary winding of transformer 1 __________

   g. Primary winding of transformer 2 __________

   h. Secondary winding of transformer 2 __________

   i. Relay 1 coil __________

   j. Relay 2 coil __________

Sequence to Complete the Lab Task

Measuring Current with a Clamp-on Ammeter

1. When you are on the job, you will need to measure the amount of current different motors and loads are drawing to be able to determine if they are working correctly. In this exercise you will be using a clamp-on ammeter to measure the amount of current of different motors and other loads. It is important to understand that the motors will have full voltage applied and they will be operating and drawing current. To operate the clamp-on ammeter, you can open its jaws and clamp them around the wire you are measuring the current draw for. You must also be aware that you will be working around open electrical terminals and moving mechanical parts such as fans and belt driven devices. You will need to take care to stay clear of moving fan blades or moving belts in belt driven equipment. Your instructor will provide the exact amount of current each device draws and you can compare your answers to these values.

   a. Current in compressor 1 run winding __________

   b. Current in compressor 2 run winding __________

c. Current in condenser fan motor 1 run winding __________

d. Current in condenser fan motor 2 run winding __________

e. Current in the electrical resistance heater of compressor __________

f. Current in the electrical resistance heating element of an electric furnace __________

2. At times you will be requested to measure very small amounts of current in a circuit. One way to get the clamp-on ammeter to measure very small amounts of current is to wrap the wire you are measuring the current draw in around the claws of the ammeter. Each time you wrap an additional coil around the claw, the amount of current is multiplied by the number of coils. For example, if you have two coils wrapped around the claw, the amount of current you are measuring will be doubled, so you will need to divide the amount being measured by two to get the exact amount of current flowing in the wire. If you have wire wrapped around the claw of the meter three times, you would need to divide the amount of current by three. In the spaces provided, indicate the amount of current measured on the meter display when you have the wire coiled around the claw. Then divide the value on the display by the number of coils to get the exact amount of current flowing in the wire. Compare your measurements with the answers your instructor has provided.

   a. Small current 1 __________
   Wrap the wire around the claw two times.

   b. Small current 1 __________
   Wrap the wire around the claw three times.

   c. Which reading gave you more accuracy? __________

   d. Small current 2 __________
   Wrap the wire around the claw two times.

   e. Small current 2 __________
   Wrap the wire around the claw three times.

   f. Which reading gave you more accuracy? __________

   g. Explain why you would wrap the wire around the claws of the ammeter.

   ____

   ____

   h. Explain when you would wrap the wire around the claw more than two times.

   ____

   ____

3. One of the things you must be aware of is, if you place the claws of the clamp-on ammeter on more than one wire, the meter will not read correctly. For example, the compressor has a wire connected to the run winding terminal and a second wire that is connected to the common terminal of the motor. If you clamped the ammeter jaws around both wires at the same time, the current in the run winding will cancel the current flowing in the wire connected to the common terminal. The result will cause the meter display to indicate zero current. Your instructor will provide a compressor motor for you to place the clamp-on ammeter claws around two wires instead of one. Since the motor is running, you can be

certain the motor is drawing current. Indicate what the clamp-on ammeter reading is when you place the claws around two wires.

____

____

4. Explain why the clamp-on ammeter would not read any current if you place it around the power cord that is plugged into a wall receptacle for a refrigerator or a window air conditioner.

____

____

Sequence to Complete the Lab Task

Measuring Voltage with an Analog and Digital Voltmeter

In this exercise you will be requested to make voltage measurements with both analog and digital voltmeters. Many technicians assume that they no longer need to use analog voltmeters with the advent of digital voltmeters. You may find that some inexpensive digital voltmeters will not be able to accurately measure voltage in some of the newer electronic circuits, and you will need to use an analog meter in some applications. You will find that it is not difficult to read the analog meter and this exercise will provide an opportunity to measure the same voltage with both a digital and an analog voltmeter and compare the readings. You may need to refer to the material in Chapter 4 that explains how to read the scale of an analog meter before you can complete this exercise and read the analog meter accurately.

You should also refer to the material in this chapter that explains that you must put the leads of your voltmeter across two terminals that have a difference of potential. For example, if you are trying to measure the voltage in a battery, you would not get a reading if you placed both meter leads on the same terminal. You must place one lead on the positive terminal and the other on the negative terminal. This is also true when you are trying to measure the voltage at the terminals of a motor or at the terminals of a fused disconnect in an AC circuit. For example, if you place the meter leads on the "run" terminal and the "start" terminal of a compressor, you would not read any voltage because they both receive voltage from line 1 (L1). You must always ensure that one of your meter leads is placed on the neutral (N) side of the circuit and then you can measure voltage to any terminal in the circuit that receives voltage from L1.

SAFETY NOTICE: It is very important that your meter is functioning properly when you are making voltage readings. Sometimes meter leads break or the meter blows a fuse and when you are trying to determine if a circuit has voltage in it, and the meter will indicate zero volts even though the circuit has full power. If you think the circuit has zero volts because the meter indicates this, you may begin to remove wires and components thinking the circuit is not powered and you could receive a severe electrical shock if you touch any of the open electrical terminals. For this reason it is very important to test your voltmeter by measuring a known voltage supply every time you take it out of your toolbox and ensure that the meter indicates the proper amount of voltage. You can determine a circuit has power by plugging in a power tool, work lamp, or other small electrical device and see if the circuit is live. If the power tool or lamp works, you can be sure the circuit has voltage, and you can place your voltmeter leads in the same terminals and check to see if it reads voltage. If the meter does not read voltage when you have the leads on a known good voltage source, you must check your meter for a bad fuse or bad terminals. This simple step will ensure that you always know whether a circuit does or does not have power turned off before you begin to work on it.

1. You should be fully supervised during this section of the lab exercise. Have your instructor initial the space provided here before you begin this section. _______

2. Your instructor will provide several operating HVAC systems and motors where you can safely take voltage measurement. It is very important to remember to work safely around equipment that has voltage applied. You should learn to place the meter in a location where you can easily read it. Next, you should take care and place the meter leads on the terminal points where you want to measure voltage and be sure to watch closely where you place the leads so as not to cause a short circuit with the metal tips of the leads. Once you have the leads safely on the terminals you are measuring, you can quickly look at the meter scale and determine the amount of voltage that is being measured. Always be aware to hold your meter leads safely on the terminals while you are looking at the meter scale. Measure the following voltages and write the value in the space provided. Compare your reading with the answers your instructor provides.

   a. Voltage across L1 and L2 in a disconnect switch __________

   b. Voltage across L1 and N in a disconnect switch __________

   c. Voltage across the terminals in an electrical wall outlet __________

   d. Voltage across the run R terminal and the common C terminal of a compressor __________

   e. Voltage across the R and C terminals of a fan motor __________

   f. Voltage across the primary winding of a control transformer __________

   g. Voltage across the secondary winding of a control transformer __________

   h. Explain why it is important to know if voltage is present in a circuit before you begin to work on the system to remove and replace a fan motor.

   ____

   ____

   i. Explain when you would use a voltmeter to accurately measure voltage at a transformer, fused disconnect, or other power supply to an HVAC system.

   ____

   ____

Checking Out

When you have completed this lab exercise, clean up your area, return all tools and supplies to their proper place, and check out with your instructor. Your instructor will initial here to indicate you are ready to check out. _______

CHAPTER 5

Symbols and Diagrams for HVAC and Refrigeration Systems

OBJECTIVES

At the end of this lab exercise you will be able to:

1. Identify common HVAC symbols.
2. Draw common HVAC symbols when given component names.
3. Identify common abbreviations for HVAC components that are shown in diagrams.
4. Look into a furnace and air-conditioning system and identify common HVAC components.

INTRODUCTION AND OVERVIEW

In this lab exercise you will be provided a number of symbols for HVAC components. When you are on the job, you will be expected to identify these common symbols that are found in electrical diagrams. You will also be requested to look at components and identify them as well as identify these common components in a working HVAC system. When you are working on these systems the instructor must ensure that the power is turned off so you can work safely around the components. You may need to refer to the material in Chapter 5 of the textbook to identify all the symbols in this section of the exercise.

TERMS

Capacitor
Coil
Compressor
Compressor contactor
Condenser
Condenser fan motor
Contactor
Control
Crankcase heater
Cutaway diagram
De-energized
Energized
Evaporator fan motor
Fan
Fuse
Fuse disconnect
Gas valve
Heater
Ladder diagram
Load
Motor
Motor starter
Normally closed
Normally open
Overload
Pressure switch
PSC motor
Relay
Schematic diagram
Single-pole switch
Single-throw switch
Solenoid
Switch
Thermostat
Transformer
Wiring diagram

MATCHING

Place the letter A–GG for the definition from the list that matches with the terms that are numbered 1–33.

Score __________

1. ____ Capacitor
2. ____ Coil
3. ____ Compressor
4. ____ Compressor contactor
5. ____ Condenser unit
6. ____ Condenser fan motor
7. ____ Contactor
8. ____ Control
9. ____ Crankcase heater
10. ____ Cutaway diagram
11. ____ De-energized
12. ____ Energized
13. ____ Evaporator fan motor
14. ____ Fan
15. ____ Fuse
16. ____ Fuse disconnect
17. ____ Gas valve
18. ____ Heater
19. ____ Ladder diagram (also called a schematic diagram)
20. ____ Load
21. ____ Motor
22. ____ Motor starter
23. ____ Normally closed
24. ____ Normally open
25. ____ Overload
26. ____ Pressure switch
27. ____ PSC motor
28. ____ Relay
29. ____ Solenoid
30. ____ Switch
31. ____ Thermostat
32. ____ Transformer
33. ____ Wiring diagram

A. The outdoor part of an air-conditioning system that houses the compressor and condensing coil.
B. An electric motor that directly drives a piston that is used to pump refrigerant in an HVAC system.
C. One or more sets of contacts that are opened or closed by a magnetic coil and is similar to a relay but normally is larger. By definition, its contacts are rated for more than 15 amps.
D. When a component is connected to a source of electrical potential (voltage).
E. A diagram of an electrical or mechanical component that shows the interior parts of the component as if someone used a saw to cut away part of the outside of the part so the interior parts can be viewed to show their location and relationship to each other.
F. An electric resistance heater that may be mounted in a cavity under the lowest part of the compressor shell called the crankcase, or it may be strapped around the base of the compressor shell so that its heat causes the oil in the compressor to warm up.
G. A device in an electrical circuit that interrupts the flow of current, such as a switch or contacts.
H. A device that can use an electrical resistance coil or a combustion chamber to provide heat to a conditioned space.

I. An electrical component that is made of two conducting plates that are separated by an insulator.

J. The act of turning an electrical component or device off. When no electrical power is applied to a component.

K. A part of a relay or solenoid that is made by tightly winding a long piece of wire into loops. When current flows through the wire, it creates a strong magnetic field.

L. An electromechanical device that converts electrical current to mechanical (rotational) motion at its shaft.

M. The fan that blows air over the evaporator coil. The fan also doubles as the furnace fan when the HVAC system is in the heating mode.

N. A relay that has contacts that can carry current in excess of 15 amps. The contactor has a coil that controls a number of normally open contacts that are wired in series with the compressor motor windings.

O. The device driven by an electric motor that moves air across the condenser coils, over evaporator coils, or through the furnace heat exchanger.

P. An industry standard for representing relay control logic. The name comes from the fact that the overall form of the diagram looks like a wooden ladder.

Q. An electrical device that consists of a single coil and one or more sets of normally open or normally closed contacts. By definition the contacts are rated for less than 15 amps.

R. A special switch that is designed to provide a disconnect for an air-conditioning system as well as provide a mounting for fuses.

S. The valve in a gas furnace that is controlled by a solenoid. When the solenoid is energized, the valve opens and allows full flow of gas to the burner.

T. A part of a motor starter that consists of a heater element and a set of normally closed contacts.

U. A switch whose contacts are activated by pressure.

V. A large relay that has a single coil and multiple sets of contacts that control voltage to a motor and a set of overload heaters and contacts that are used to protect the motor against overcurrent.

W. An electrical device that uses current and converts it to heat, light, and motion, such as a resistance heating coil, magnetic coil, motor, solenoid, or indicator lamp.

X. A condition with switch contacts or relay contacts where they have high resistance. A normal condition is considered when no power is applied to the circuit.

Y. Permanent split-capacitor motor.

Z. An electrical diagram that shows the location of all the electrical components in the circuit and the wires that connects to them.

AA. An electrical device with two coils (primary winding and secondary winding) that are located in close proximity. Voltage is applied to the primary winding and voltage is taken off the secondary winding.

BB. A coil of wire that can control a valve or other device that requires linear motion. The coil becomes a very strong magnet when current is supplied to it.

CC. An electrical device with one or more sets of contacts.

DD. A temperature-controlled switch that controls the furnace, air conditioner, and fan motors.

EE. A condition with switch contacts or relay contacts where they have low resistance. A normal condition is considered when no power is applied to the circuit.

FF. A device designed as a one-time protection against overcurrent or short-circuit current. The fusible link melts and causes an open circuit when its current rating is exceeded.

GG. The motor that turns the fan that moves air over the condensing coil.

TRUE OR FALSE

Place a *T* or *F* in the blank to indicate if the statement is true or false.

Score __________

1. ____ A wiring diagram shows the relative location of each component in a circuit.
2. ____ An electrical load is a device that has resistance and uses current.
3. ____ A schematic (ladder) diagram shows the relative location of each component in a circuit.
4. ____ A furnace fan can also be the condenser fan in some systems.
5. ____ The thermostat for a furnace is an electrical load.

MULTIPLE CHOICE

Circle the letter that represents the correct answer to each question.

Score __________

1. The control in the furnace that protects the furnace from overheating and causing a fire is the:
   a. fan switch.
   b. high-limit switch.
   c. fan relay.

2. The "operator" of an electrical switch is the part of the switch that:
   a. causes the action that makes the switch change from open to closed or closed to open.
   b. conducts current.
   c. is the same as the contacts.

3. The auto/on switch on the thermostat allows the fan to operate continuously or with the:
   a. gas valve.
   b. transformer.
   c. compressor.

4. The thermostat terminals are identified with the letters R, Y, W, and G, which stand for:
   a. the colors of the thermostat wires (red, yellow, white, and green).
   b. the operation of the thermostat (auto, manual, and semiautomatic).
   c. the primary and secondary windings on the control transformer.

5. The fan and limit switch in the gas furnace control the temperature at which the fan turns on and the:
   a. temperature at which the gas is ignited.
   b. temperature at which the air conditioner turns off.
   c. maximum temperature the furnace can reach before power to the gas valve is turned off.

LAB EXERCISE: UNDERSTANDING COMPONENTS AND THEIR ELECTRICAL SYMBOLS

Safety for this Lab Exercise

You will not be working around electrical power in this exercise. The activities for this lab use paper and pencil as well as some components that will be displayed on a tabletop. You will be requested to look into a furnace and an HVAC system to identify components, but the power will be turned off and secured, so there are no safety concerns for this lab exercise.

Tools and Materials Needed to Complete the Lab Exercise

Your instructor will need to set up a number of HVAC components on a tabletop for you to identify and draw their electrical symbol. These components should not be connected to power and can be placed on a table so students can pick them up and handle them in order to get a closer look. You will also need to secure a furnace and an HVAC system, which can be a packaged unit or a split system. The power to this system must be turned off and locked out so that no electrical threat is posed. It is essential that the instructor ensure that the power is turned off so that fans do not run and no electrical shock hazard exists. Your instructor should gather the following components and put them on display for you to identify.

1. Relay
2. Thermostat
3. Low-pressure switch
4. Condenser fan with motor
5. Compressor
6. evaporator fan with motor
7. Run capacitor
8. Start capacitor
9. Transformer
10. Crankcase heater
11. Normally closed push-button switch
12. Normally open push-button switch
13. Single-pole toggle switch

References to the Text

Refer to Chapter 5 in the textbook for additional information. You may need to read sections of the chapter again to help you understand the material in this exercise.

Sequence to Complete the Lab Task

Identifying Electrical Symbols

1. Write the name of each symbol in the space directly below it.

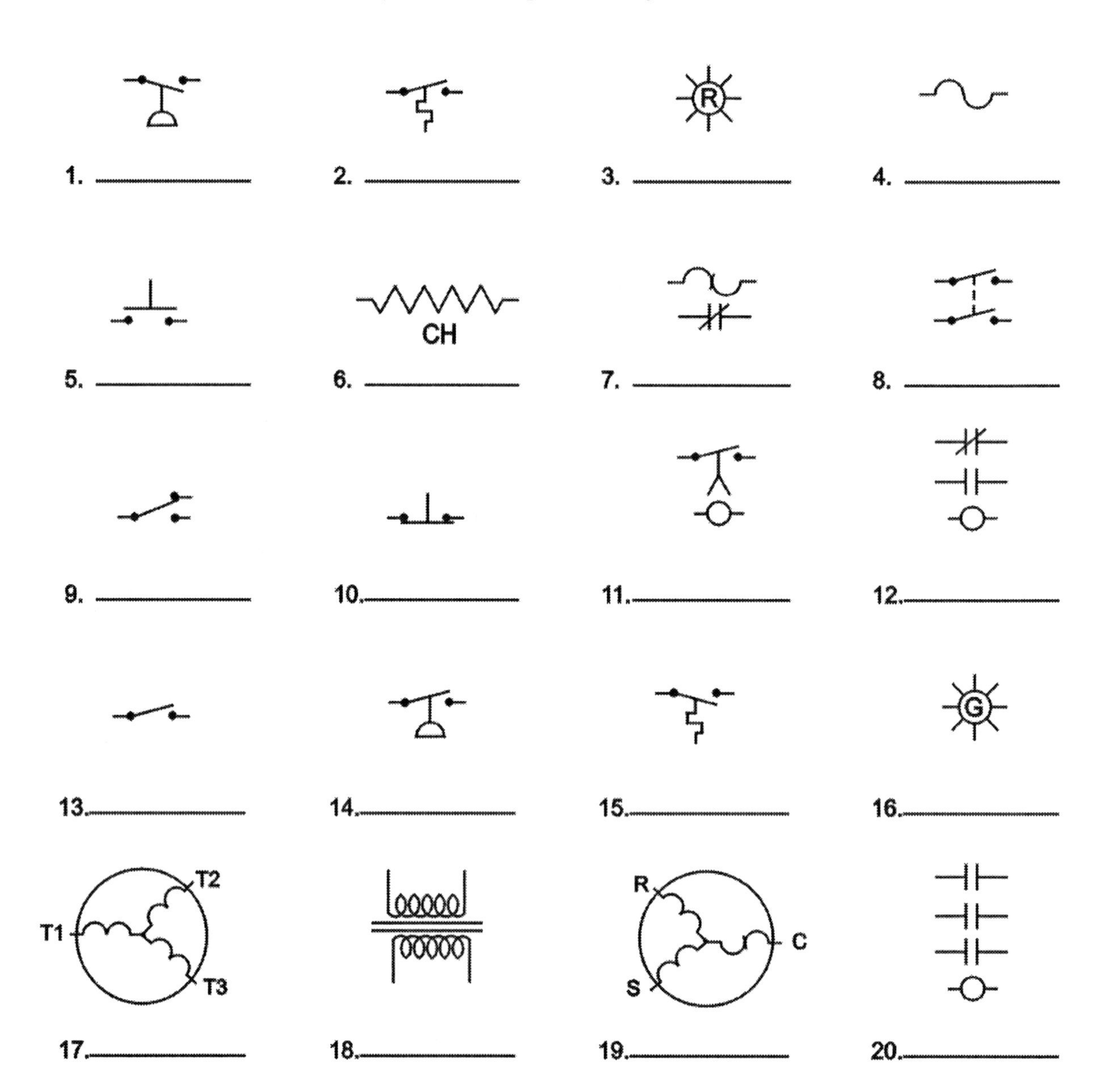

Sequence to Complete the Lab Task

Drawing Symbols for the Following Symbols

1. Your instructor will provide a number of components that you can pick up and inspect to ensure that you are identifying them correctly. Your instructor will place a tag with the letters A–M on each component that you are to identify. When you have identified a component, put the letter in the space next to the component name. In the additional space to the right of the component name, draw the symbol for the component. Have your instructor review your list and the symbol for each component for accuracy.

____ Relay

____ Thermostat

____ Low-pressure switch

____ Condenser fan with motor

____ Compressor

____ Evaporator fan with motor

____ Run capacitor

____ Start capacitor

____ Transformer

____ Crankcase heater

____ Normally closed push-button switch

____ Normally open push-button switch

____ Single-pole toggle switch

Sequence to Complete the Lab Task

Identifying Electrical Components in an Air-Conditioning System

1. Your instructor will provide a packaged or split-system air conditioner. Your task for this exercise is to identify the following components. Your instructor will place a tag with the letters A–J on each component you are to identify. You should place the letter for the component next to the name of the component provided in the list. Have your instructor review your list for accuracy.

____ Compressor relay

____ Thermostat

____ Evaporator fan with motor

____ Run capacitor

____ Start capacitor

____ Condenser fan with motor

____ Compressor

____ Transformer

____ Crankcase heater

____ Disconnect switch

Sequence to Complete the Lab Task

Identifying Electrical Components in a Gas Furnace

1. Your instructor will provide a gas furnace, and will place a tag with the letters A–D on each component that you are to identify. Your task for this exercise is to identify the following components. You should place the letter for the component next to the name of the component provided in the list. Have your instructor review your list of these components for accuracy.

____ Thermostat ____ Gas valve

____ Furnace fan with motor ____ Transformer

Sequence to Complete the Lab Task

Identifying Abbreviations of Electrical Components

1. When you are working with electrical diagrams, many of the components will have an abbreviation that is associated with them in the diagram. You will be expected to know the abbreviation, and in this exercise you will write the name of the component next to the abbreviation. You may need to refer to the material about abbreviations in the textbook.

ALS ______________________________ IDM ______________________________

BM ______________________________ IDR ______________________________

BMR ______________________________ LED ______________________________

CAP ______________________________ LEVS ______________________________

COM ______________________________ LGPS ______________________________

CPU ______________________________ LS ______________________________

DPDT ______________________________ NC ______________________________

DPST ______________________________ NO ______________________________

DSS ______________________________ OL ______________________________

EAC ______________________________ PCB ______________________________

FRS ______________________________ PRI ______________________________

FS ______________________________ PS ______________________________

FU ______________________________ SEC ______________________________

GV ______________________________ SPDT ______________________________

GVR ______________________________ SPST ______________________________

HIS ______________________________ TRAN ______________________________

HUM ______________________________

Checking Out

When you have completed this lab exercise, clean up your area, return all tools and supplies to their proper place, and check out with your instructor. Your instructor will initial here to indicate you are ready to check out. _______

CHAPTER 6

Reading and Writing Schematic (Ladder) and Wiring Diagrams

OBJECTIVES

At the end of this lab exercise you will be able to:

1. Explain the difference between a schematic (ladder) diagram and a wiring diagram.
2. Create a schematic (ladder) diagram from a wiring diagram.
3. Create a wiring diagram from a schematic (ladder) diagram.
4. Read a schematic (ladder) diagram.
5. Read a wiring diagram.

INTRODUCTION AND OVERVIEW

When you are on the job troubleshooting and repairing HVAC systems, you will be expected to read and draw electrical wiring and schematic (ladder) diagrams. One of the challenges you will face as an HVAC technician is that you will have a wiring diagram and you will need to convert it into a schematic diagram, or you may have a schematic diagram and you may need to convert it into a wiring diagram. In this lab exercise you will be provided a wiring diagram and schematic (ladder) diagram of an HVAC system and you will be requested to convert a wiring diagram into a schematic, and a schematic diagram into a wiring diagram. You may need to refer to the material in Chapter 6 of the textbook in order to complete these exercises.

TERMS

Capacitor
Compressor
Compressor contactor
Condenser fan
Electrical control
Electrical load
Fan relay
HVAC electrical symbols
Legend
Schematic (ladder) diagram
Thermostat
Transformer
Wiring diagram

MATCHING

Place the letter A–M for the definition from the list that matches with the terms that are numbered 1–13.

Score __________

1. ____ Capacitor
2. ____ Compressor
3. ____ Compressor contactor
4. ____ Condenser fan motor
5. ____ Electrical control
6. ____ Electrical load
7. ____ Fan relay
8. ____ Legend
9. ____ HVAC electrical symbols
10. ____ Schematic (ladder) diagram
11. ____ Transformer
12. ____ Thermostat
13. ____ Wiring diagram

A. Electrical symbols that represent fan motors and compressor motors, electrical heating coils, and a wide variety of controls including pressure switches, flow switches, temperature switches, and level switches.

B. A part of a drawing or diagram that explains special symbols and conventions used in the drawing.

C. The devices in an electrical system that consume energy and convert it to motion, heat, or light.

D. An electrical component that is made of two conducting plates that are separated by an insulator.

E. The fan on a condenser unit that moves air over the condenser coils.

F. An electric motor that directly drives a piston that is used to pump refrigerant in an HVAC or refrigeration system.

G. A device that has a coil and one or more contacts that can carry current in excess of 15 amps. It has a coil that controls a number of normally open contacts that are wired in series with the compressor motor windings.

H. An electrical diagram that shows the location of all the electrical components in the circuit and the wires that connect to them.

I. A diagram that is designed to show the sequence of operation of an electrical control system.

J. The relay that consists of a coil and one or more sets of contacts in the HVAC system that controls the furnace fan. The coil of the relay is connected to the G terminal on the thermostat, and the contacts of the relay are connected in series with the fan motor.

K. The switches in a schematic or wiring diagram that provide the sequence turn loads on or off.

L. A temperature-controlled switch that controls the furnace, air conditioner, and fan motors. The thermostat can be heating only, cooling only, or heat/cool.

M. An electrical device with two coils (primary winding and secondary winding) that are located in close proximity. Voltage is applied to the primary winding and voltage is taken off the secondary winding.

TRUE OR FALSE

Place a *T* or *F* in the blank to indicate if the statement is true or false.

Score __________

1. ____ The schematic (ladder) diagram shows the sequence of operation for a system.
2. ____ A control is a device in an electrical system that uses current and produces work.
3. ____ The loads in a schematic (ladder) diagram are usually located in the top part of the diagram.
4. ____ The relay coil for the evaporator fan relay is located in the control part of the circuit, and the contacts for the relay are located in the load part of the circuit.
5. ____ The main difference between a packaged unit and a split-system air conditioner is that the evaporator fan of the split system is usually mounted in the indoor unit.

MULTIPLE CHOICE

Circle the letter that represents the correct answer to each question.

Score __________

1. The compressor motor in the air-conditioning system electrical diagram is located in:
   a. the load circuit.
   b. the control circuit.
   c. both circuits depending on the type of system.

2. The condenser fan motor is usually controlled by the:
   a. same controls that control the compressor.
   b. same controls that control the evaporator fan.
   c. high-pressure switch.

3. In the refrigeration system, the compressor, condenser fan, and evaporator fan:
   a. are each controlled by separate controls.
   b. are all turned on and off by the thermostat.
   c. can be turned off by the defrost switch or the high-pressure switch.
   d. All the above
   e. Only a and b

4. In the packaged air-conditioning system, the evaporator fan:
   a. is located in the indoor unit (furnace).
   b. is located in the same unit as the compressor.
   c. could be located in the indoor unit or the outdoor unit.
   d. All the above
   e. Only a and b

5. In the window air conditioner, the:
   a. same motor that has a dual shaft that sticks out each end of the motor usually powers the evaporator fan and condenser fan.
   b. multispeed switch on the front panel controls the evaporator fan.
   c. evaporator fan is controlled by a separate switch than the condenser fan.
   d. All the above
   e. Only a and b

LAB EXERCISE: CONVERTING A WIRING DIAGRAM TO A SCHEMATIC (LADDER) DIAGRAM

Safety for this Lab Exercise

You will not be working around electrical power in this exercise. The activities for this lab use paper and pencil and you will not be near anything that poses any safety concerns.

Tools and Materials Needed to Complete the Lab Exercise

You will be provided a wiring diagram and a schematic (ladder) diagram of an HVAC system in this exercise. You will have space provided in this exercise to draw the converted diagrams.

References to the Text

Refer to Chapter 6 in the textbook for additional information. You may need to read sections of the chapter again to help you understand the material in this exercise.

Sequence to Complete the Lab Task

Converting a Wiring Diagram to a Schematic (Ladder) Diagram

1. The wiring diagram provided in Figure 6–1 is for an air-conditioning system. In the space provided below, convert this wiring diagram into a schematic (ladder) diagram. Have your instructor review your work when you have completed the drawing.

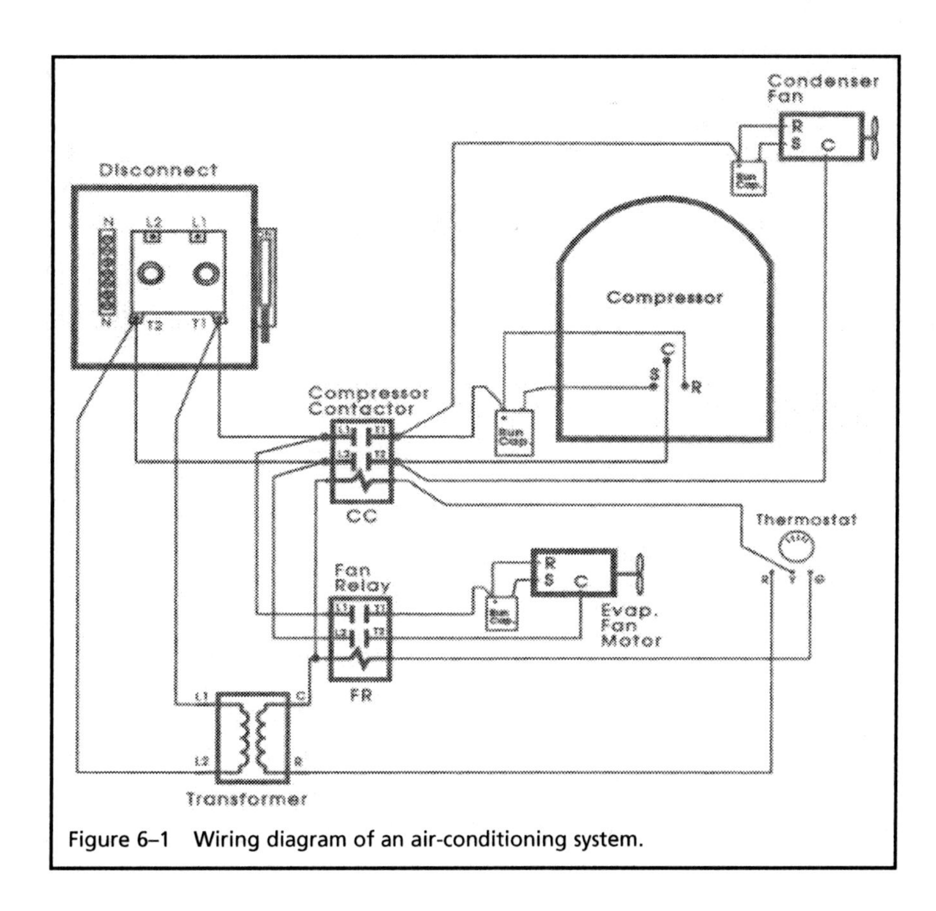

Figure 6–1 Wiring diagram of an air-conditioning system.

Sequence to Complete the Lab Task

Converting a Schematic (Ladder) Diagram into a Wiring Diagram

1. The schematic diagram provided in Figure 6–2 is for an air-conditioning system. In the space provided below, convert this schematic (ladder) diagram into a wiring diagram. Have your instructor review your work when you have completed the drawing.

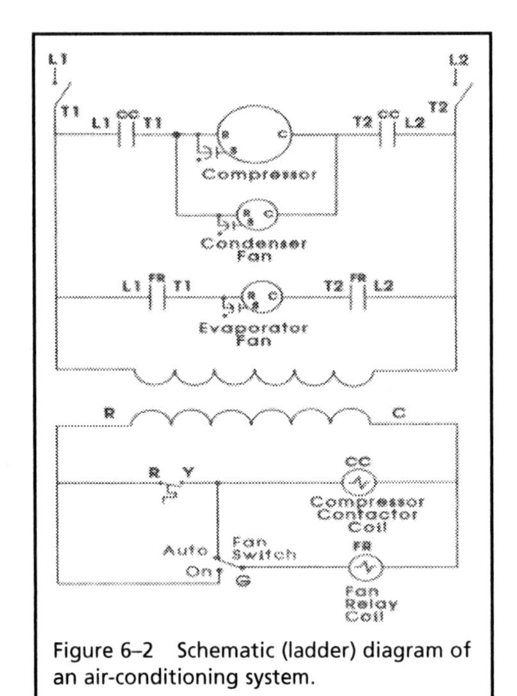

Figure 6–2 Schematic (ladder) diagram of an air-conditioning system.

Sequence to Complete the Lab Task

Creating a Schematic (Ladder) Diagram by Looking at Components in an HVAC System

1. At times you may need to create a schematic (ladder) diagram or wiring diagram by just looking at the components in an HVAC system and following the wires to see where they are connected. Your instructor will provide a packaged or split-system air conditioner that has the power turned off. Your task is to identify the components in the electrical system, follow the wires, and create a wiring diagram or a schematic (ladder) diagram. Draw your diagram in the space provided. Have your instructor review it for accuracy.

Sequence to Complete the Lab Task

Identifying Electrical Components in a Gas Furnace

1. Your instructor will provide a gas furnace. Your instructor will place a tag with the letters A–D on each component in the furnace. Your task for this exercise is to identify the following components. You should place the letter for the component next to the name of the component provided in the list. Have your instructor review your list of these components for accuracy.

____ Thermostat	____ Furnace fan with motor
____ Gas valve	____ Transformer

Sequence to Complete the Lab Task

Understanding the System Operation of an HVAC System by Reading a Diagram

1. One of the tasks you will be expected to do when you are on the job is to read a schematic diagram and be able to explain what is happening in the circuit. Answer the following questions about the circuit shown in Figure 6–3. Have your instructor review your answers for accuracy.

   a. What coil must be energized to get voltage to the compressor motor?

   ____

   ____

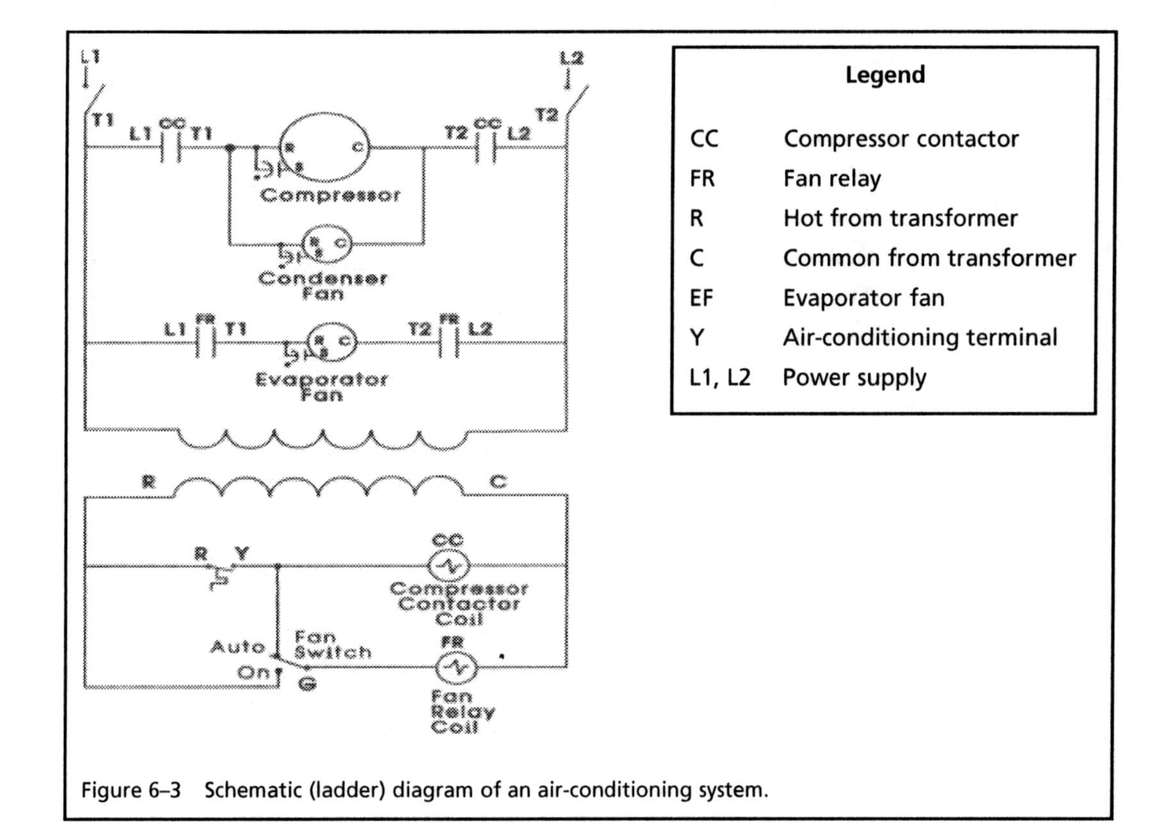

Figure 6–3 Schematic (ladder) diagram of an air-conditioning system.

b. What must occur in the system to get the evaporator fan to turn on?

____

____

c. When will the condenser fan run?

____

____

Sequence to Complete the Lab Task

Reading Wiring Diagrams and Schematic (Ladder) Diagrams for the Outdoor Section of a Split-System Air Conditioner

Answer the following questions, referring to Figure 6–4.

1. When is the crankcase heater energized?
   a. When the compressor contactor is energized
   b. Any time power is applied to the system
   c. Only when the compressor contactor is not energized

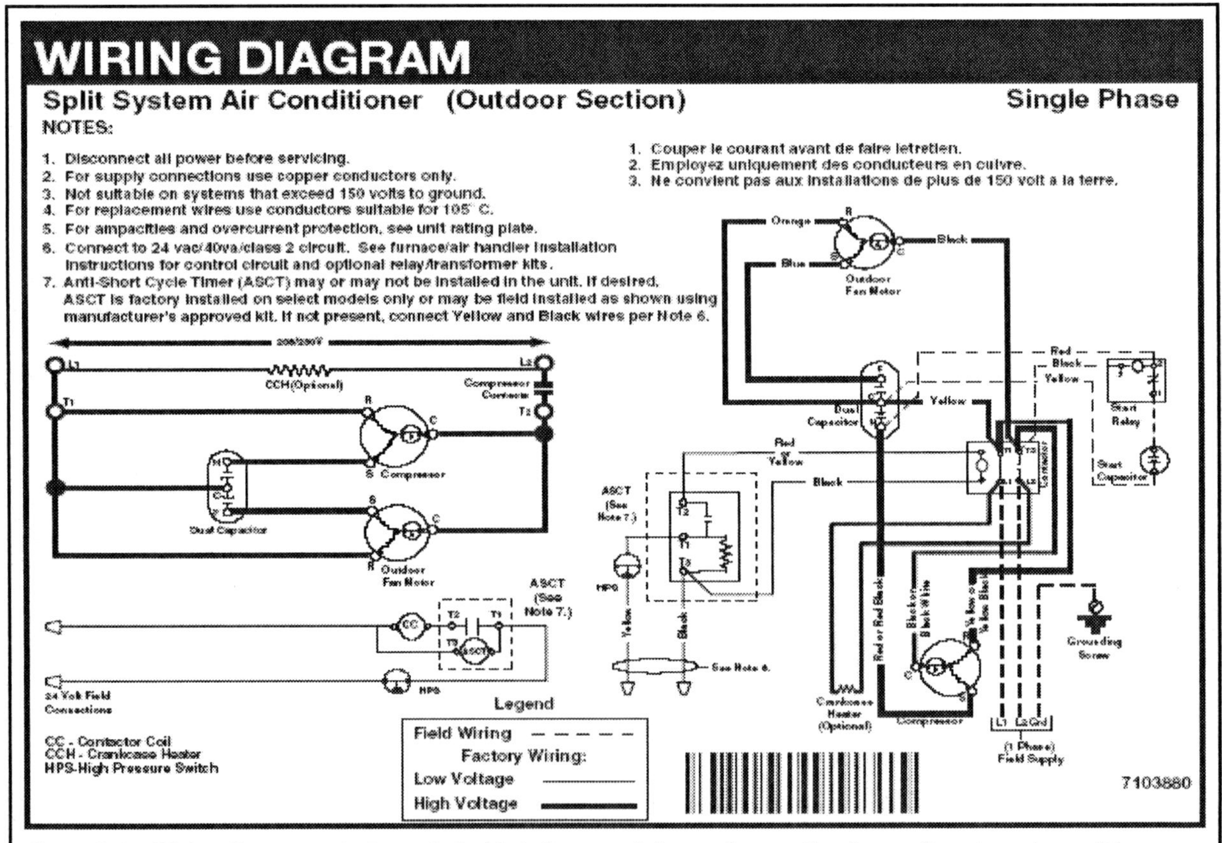

Figure 6–4 Wiring diagram and schematic (ladder) diagram of the outdoor section for a split-system air conditioner. (*Courtesy of Nordyne*)

2. What causes the compressor to become energized?
   a. When the compressor contactor is energized
   b. Any time power is applied to the system
   c. Only when the compressor contactor is not energized

3. What motor does the capacitor help start?
   a. Only the compressor
   b. Only the condenser fan motor
   c. Both the compressor and condenser fan motor
   d. None of the motors in this system use a capacitor.

4. What happens if the high-pressure switch (HPS) opens?
   a. The compressor contactor coil is de-energized.
   b. The crankcase heater is de-energized.
   c. Only the compressor is de-energized.
   d. It is impossible to tell.

Sequence to Complete the Lab Task

Reading Wiring Diagram and Schematic (Ladder) Diagram for a Packaged Air Conditioner

Answer the following questions, referring to Figure 6–5.

1. What is the size of the fuse that is protecting the secondary side of the control transformer?
   a. 2 amps
   b. 3 amps
   c. 4 amps
   d. Cannot tell from this diagram

2. What controls voltage to energize the compressor motor and condenser fan motor?
   a. Compressor contactor
   b. Blower relay
   c. The start capacitor

3. What controls voltage to energize the evaporator (blower) motor?
   a. Compressor contactor
   b. Blower relay
   c. The start capacitor

4. The dual capacitor is connected to the:
   a. blower motor and the compressor.
   b. blower motor and the condenser fan motor.
   c. compressor and the condenser fan motor.

5. If the pressure switch opens due to high pressure:
   a. the compressor contactor coil will not be energized.
   b. no voltage will be supplied to the transformer primary winding.
   c. the coil for the blower motor will not be energized.
   d. All of the above

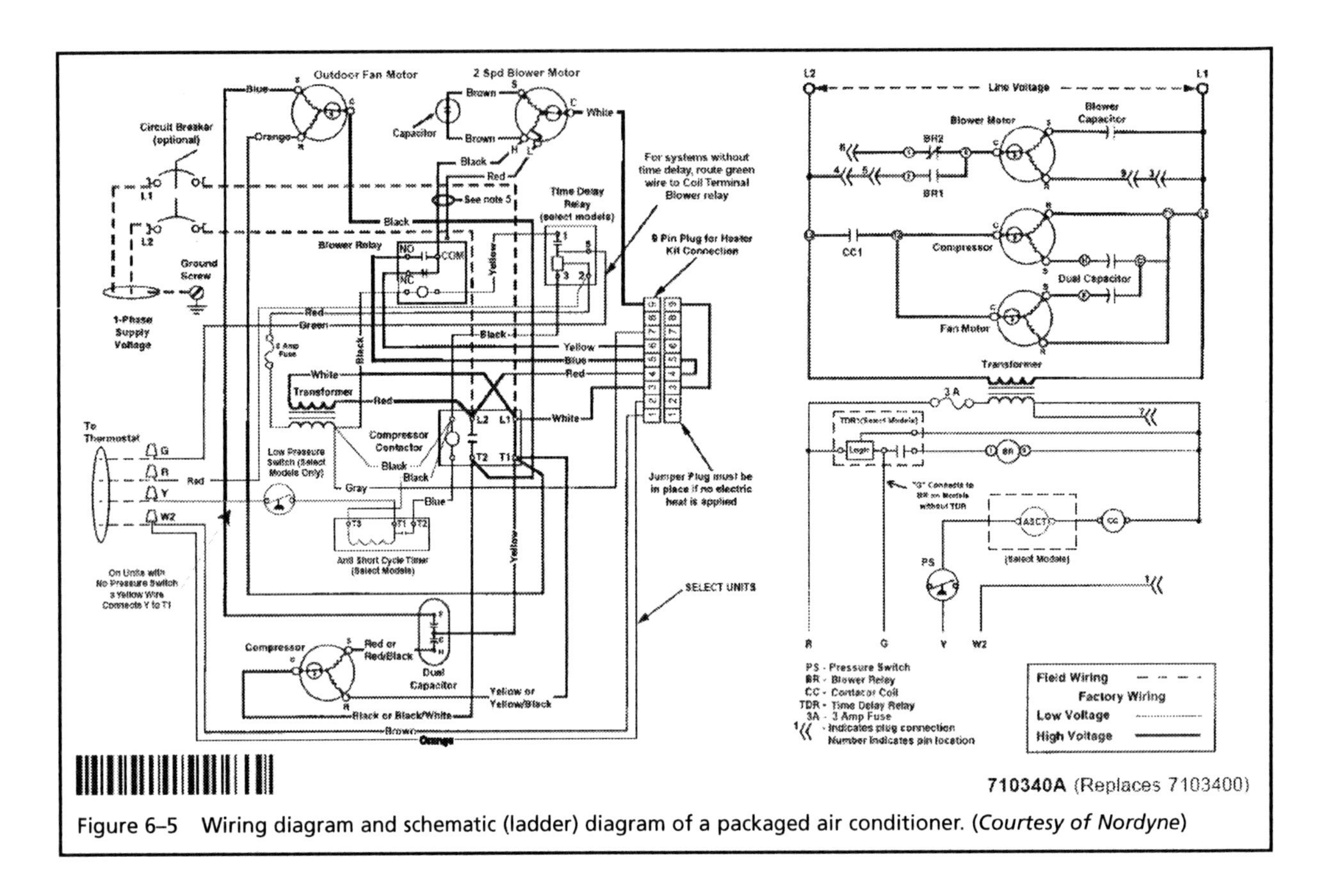

Figure 6–5 Wiring diagram and schematic (ladder) diagram of a packaged air conditioner. (*Courtesy of Nordyne*)

6. The coil of the blower motor relay receives voltage from the:
   a. G terminal of the thermostat.
   b. Y terminal of the thermostat.
   c. W2 terminal of the thermostat.
   d. All of the above

7. The coil of the compressor contactor receives voltage from the:
   a. G terminal of the thermostat.
   b. Y terminal of the thermostat.
   c. W2 terminal of the thermostat.
   d. All of the above

Sequence to Complete the Lab Task

Reading Wiring Diagrams and Schematic (Ladder) Diagrams for a Packaged Heat Pump

Answer the following questions, referring to Figure 6–6.

1. What is the size of the fuse that is protecting the secondary side of the control transformer?
   a. 2 amps
   b. 3 amps
   c. 4 amps
   d. Cannot tell from this diagram

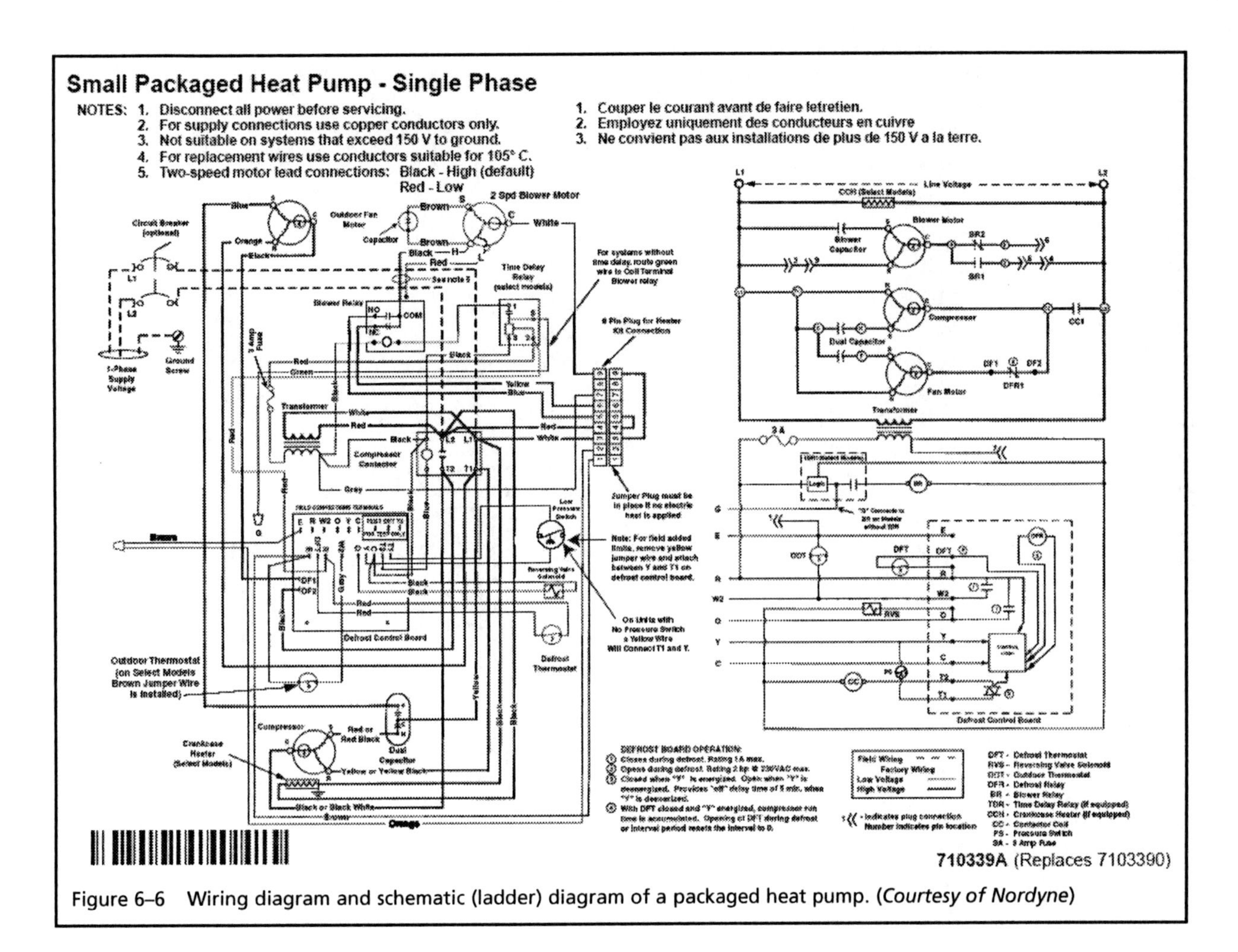

Figure 6–6 Wiring diagram and schematic (ladder) diagram of a packaged heat pump. (*Courtesy of Nordyne*)

2. What controls voltage to energize the compressor motor and condenser fan motor?
   a. Compressor contactor
   b. Blower relay
   c. The start capacitor

3. What controls voltage to energize the evaporator (blower) motor?
   a. Compressor contactor
   b. Blower relay
   c. The start capacitor

4. The dual capacitor is connected to the:
   a. blower motor and the compressor.
   b. blower motor and the condenser fan motor.
   c. compressor and the condenser fan motor.

5. The coil of the relay for the blower motor receives voltage from the:
   a. G terminal of the thermostat.
   b. Y terminal of the thermostat.
   c. W2 terminal of the thermostat.
   d. All of the above

6. The coil of the compressor contactor (CC) receives voltage from the:
   a. G terminal of the thermostat.
   b. Y terminal of the thermostat.
   c. W2 terminal of the thermostat.
   d. T2 terminal of the thermostat.

7. The condenser fan motor will:
   a. always run any time the compressor motor is running.
   b. be turned off by the normally closed contacts of the DFR (defrost relay).
   c. run any time the indoor (blower) motor is running.
   d. It is impossible to tell from this diagram.

8. The crankcase heater (CCH):
   a. only runs when the compressor motor is running.
   b. only runs when the blower motor is running
   c. is energized at all times when voltage is applied to the system.
   d. It is impossible to tell from this diagram.

9. The reversing valve solenoid receives voltage from the:
   a. G terminal of the thermostat.
   b. Y terminal of the thermostat.
   c. W2 terminal of the thermostat.
   d. O terminal of the thermostat.

Sequence to Complete the Lab Task

Reading a Schematic (Ladder) Diagram for an Electric Furnace

Answer the following questions, referring to Figure 6–7.

1. The coil for the fan relay receives voltage from the:
   a. G terminal of the thermostat.
   b. W terminal of the thermostat.
   c. contacts of SEQ1.
   d. contacts of SEQ2.

2. The coil of Sequencer 1 (SEQ1) receives voltage from the:
   a. G terminal of the thermostat.
   b. W terminal of the thermostat.
   c. contacts of SEQ2.
   d. All of the above

3. The coil of Sequencer 2 (SEQ2) receives voltage from the:
   a. G terminal of the thermostat.
   b. W terminal of the thermostat.
   c. contacts of SEQ1.
   d. All of the above

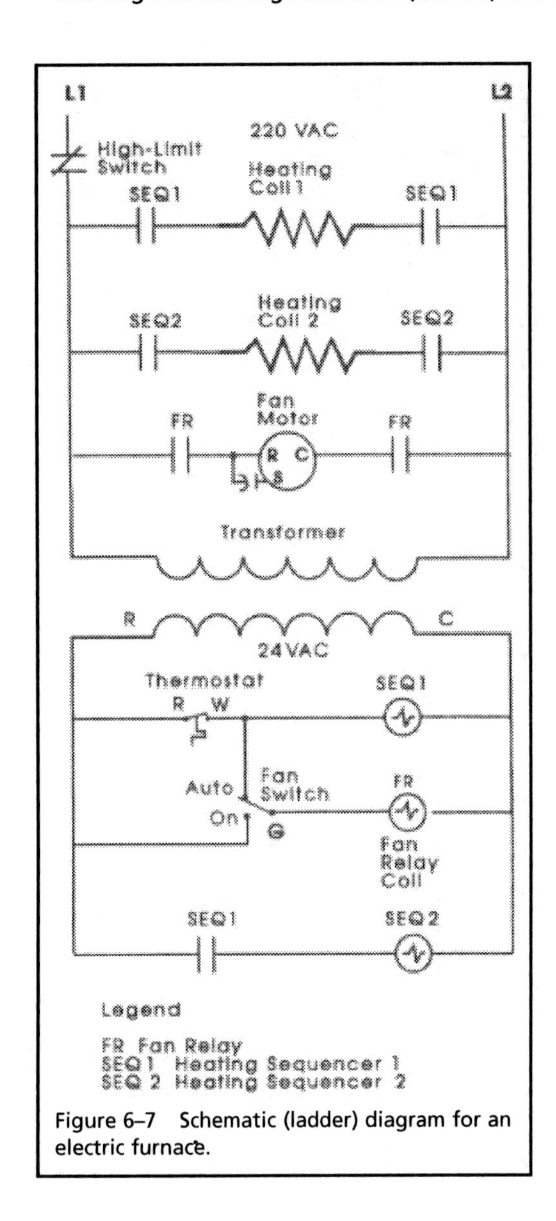

Figure 6–7 Schematic (ladder) diagram for an electric furnace.

4. Heating coil 1 receives voltage from the contacts of:
   a. SEQ1.
   b. SEQ2.
   c. FR.
   d. All of the above

5. Heating coil 2 receives voltage from the contacts of:
   a. SEQ1.
   b. SEQ2.
   c. FR.
   d. All of the above

Sequence to Complete the Lab Task

Reading Wiring Diagrams and Schematic (Ladder) Diagrams for an Oil Furnace

Answer the following questions, referring to Figure 6–8.

1. What is the voltage supply for the system?
   a. 24 VAC
   b. 120 VAC
   c. 240 VAC
   d. Cannot tell from this diagram

2. What happens if the limit switch opens due to high temperature?
   a. Power is turned off to the primary control.
   b. Power is turned off to the blower motor.
   c. Power is turned off to the cooling contactor.

3. The flame sensor is connected to terminals:
   a. T and T on the primary control.
   b. F and F on the primary control.
   c. R and W on the thermostat.

4. How many speeds does the blower motor have?
   a. Two: high and low
   b. Three: high, medium, and low
   c. Four: high, medium high, medium low, and low
   d. It is impossible to tell from this diagram.

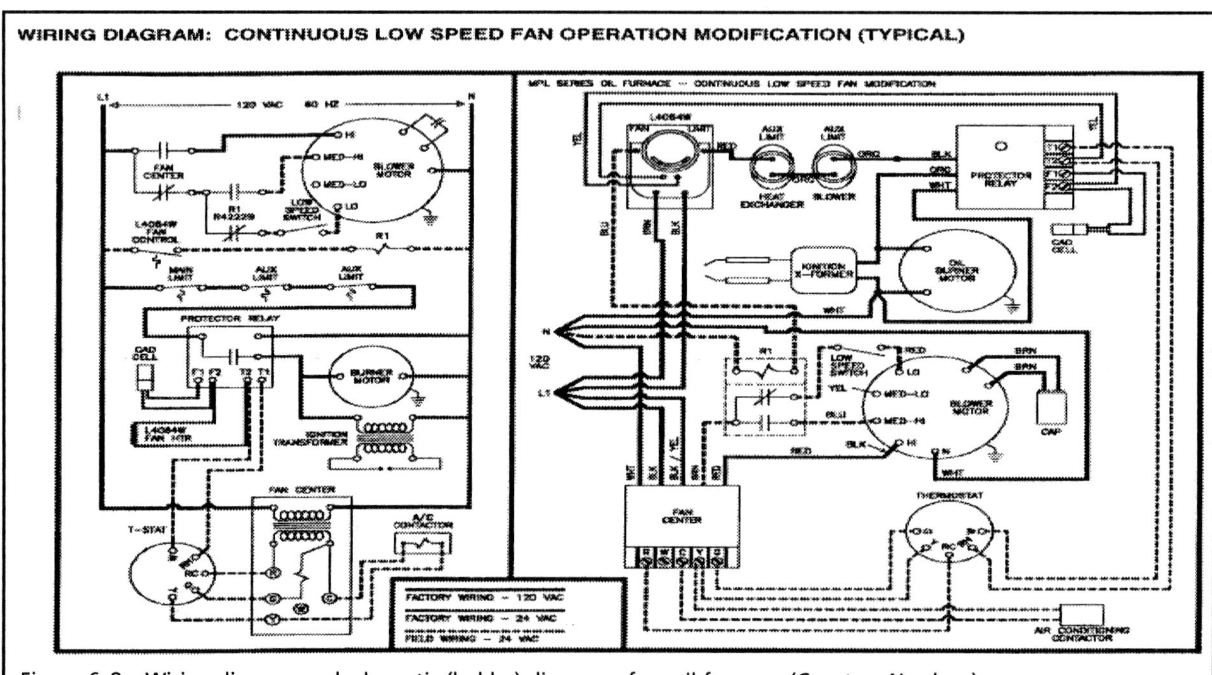

Figure 6–8 Wiring diagram and schematic (ladder) diagram of an oil furnace. (*Courtesy Nordyne*)

Sequence to Complete the Lab Task

Reading A Schematic (Ladder) Diagram for a Refrigeration Case

Answer the following questions, referring to Figure 6–9.

1. What happens if one of the doors of the refrigeration case is opened?
   a. The compressor is turned off.
   b. The condenser fan is turned off.
   c. The evaporator fan is turned off.

2. What happens when the high-pressure cutout switch opens?
   a. Only the compressor is turned off.
   b. Only the condenser fan is turned off.
   c. The evaporator fan is turned off
   d. Both the compressor and the condenser fan are turned off.

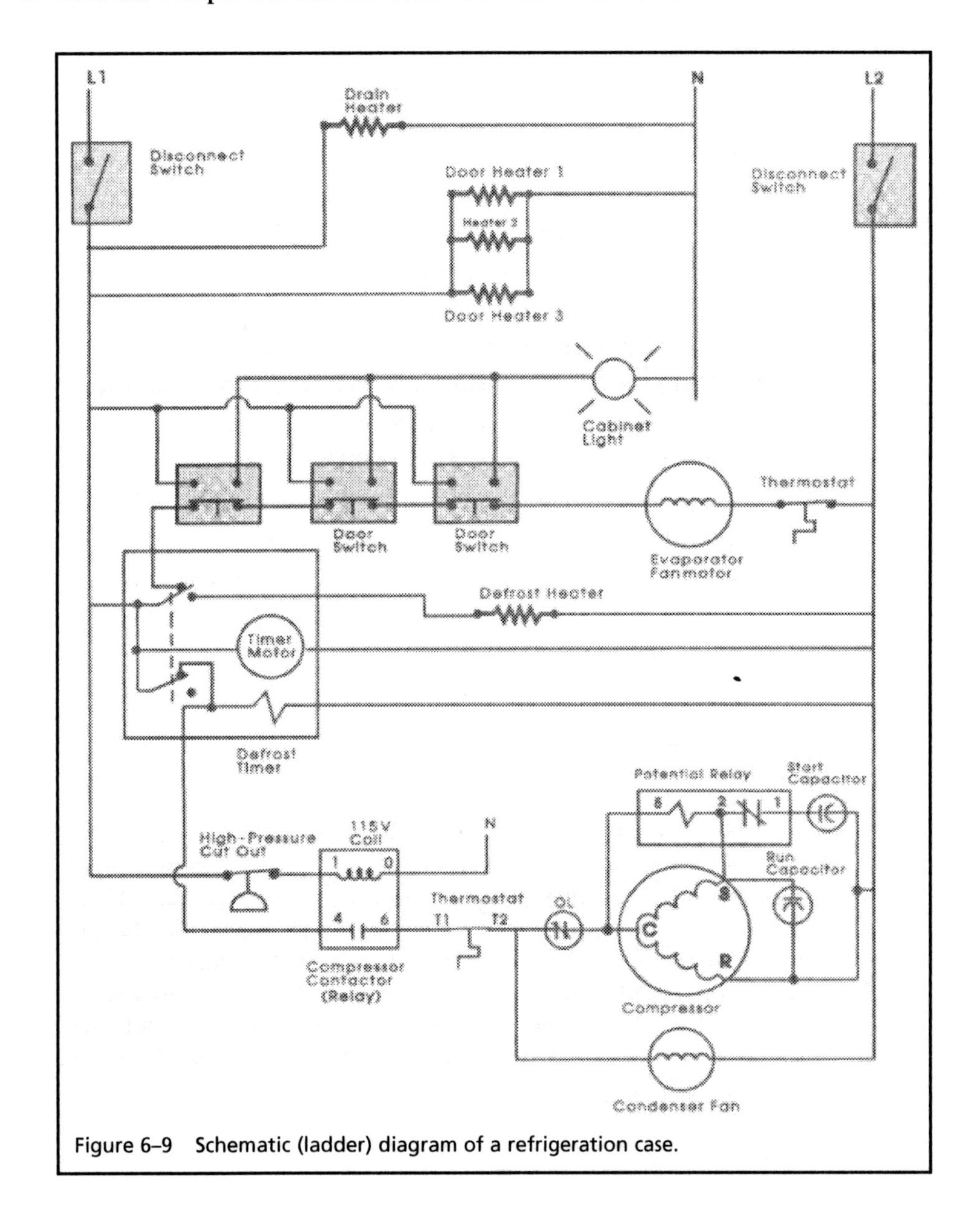

Figure 6–9 Schematic (ladder) diagram of a refrigeration case.

3. What are the components that help the compressor start?
    a. The start capacitor
    b. The run capacitor
    c. The potential relay
    d. All the above
    e. Only the start and run capacitors

4. What turns the defrost heating coil on and off?
    a. The thermostat
    b. The door switches
    c. The defrost switch
    d. All the above

Checking Out

When you have completed this lab exercise, clean up your area, return all tools and supplies to their proper place, and check out with your instructor. Your instructor will initial here to indicate you are ready to check out. _______

CHAPTER 7

Magnetism, Alternating Current, and Power Distribution

OBJECTIVES

At the end of this lab exercise you will be able to:

1. Describe a magnet and explain how it works.
2. Explain the difference between a permanent magnet and an electromagnet.
3. Explain the term alternating current (AC).
4. Explain the operation of a fused disconnect switch and how to safely test for voltage at the line side and the load side.

INTRODUCTION AND OVERVIEW

The theory of operation for all types of transformers, motors, and relay coils can be explained with several simple magnetic theories. As a technician who works on air-conditioning, heating, and refrigeration systems, you will need to comprehend fully all magnetic theories so that you will understand how these components operate. You must understand how a component is supposed to operate before you can troubleshoot it, and perform tests to determine if it has failed. Understanding magnetic theory will make this job easier. This exercise will have a number of questions to help you review the information about magnets and how they are used in motors, transformers, and other coils.

Alternating current (AC) is used to provide power to residential and commercial air-conditioning and refrigeration systems. The electrons in alternating current flow in one direction for the first half cycle, and then flows in the other direction during the second half cycle. In the United States, the frequency for alternating current is 60 cycles per second (60 hertz). As a technician you will be involved in the power distribution of the AC voltage from the main circuit breaker box to the air conditioners and refrigeration systems. The point where voltage comes into the HVAC and refrigeration equipment is the disconnect switch. If the disconnect switch has fuses in it to protect the system, it is called a fused disconnect. When you are troubleshooting an HVAC or refrigeration system that does not have voltage supplied to it, you will need to trace the voltage to the source at the circuit breaker panel, and then to the disconnect switch, and test it at each point. If you find the fuses in the disconnect are blown, you will need to do further troubleshooting to determine the cause of the overcurrent. The disconnect switch is also used to turn off all power to a system so you can safely work on it to remove components and replace them with new ones. You may also be asked to "lock out" the disconnect switch by placing a padlock on the switch when it is in the off position. You will need to become accurate in making voltage tests in the disconnect switch when it is in the off position to ensure that voltage has turned off completely and it is safe to test fuses or work on the equipment.

When AC voltage is applied to coils (inductors) in motors, transformers, and solenoids, the circuit will get increased opposition to the flow of current, caused by inductive reactance. *Inductive reactance* is similar to resistance in a direct current (DC) circuit in that it helps keep current lower as it increases. When capacitors are used to help start single-phase AC motors, the capacitors also cause an opposition to current flow, which causes *capacitive reactance*. When a circuit has inductive reactance or capacitive reactance, they form an opposition called impedance. *Impedance* is the total opposition in an AC circuit that includes opposition from resistance, inductive reactance, and capacitive reactance. As a technician, you will not measure inductive reactance or capacitive reactance; rather, you will see the result of their opposition to current flow in the amount of current that is in a circuit.

Another important task you will need to learn is the function and operation of the fused disconnect switch. The disconnect switch can have fuses, or it can be just a switch. As a technician, you will need to understand the operation of the disconnect switch and how to test for voltage inside it on the line-side terminals, which are located at the top of the disconnect and on the load-side terminals, which are located on the bottom of the switch. You will be requested to test the disconnect switch for voltage in this exercise and use this information for future lab exercises.

TERMS

Alternator
Apparent power
Capacitive reactance
Control transformer
Delta-connected transformer
Frequency
Fuse disconnect
Hertz
Impedance
Inductive reactance
Neutral
Peak-to-peak voltage
Peak voltage
Period
Phase
Power factor
Primary winding
Reactance
RMS voltage
Root mean square (rms)
Secondary winding
Sine wave
Single phase
Step-down transformer
Step-up transformer
Transformer
True power
Turns ratio
VA rating for transformer
Wye-connected transformer

MATCHING

Place the letter A–T for the definition from the list that matches with the terms that are numbered 1–20.

Score __________

1. ____ Alternating current (AC)
2. ____ Apparent power
3. ____ Capacitive reactance
4. ____ Disconnect (switch)
5. ____ Frequency
6. ____ Hertz
7. ____ Impedance
8. ____ Inductive reactance
9. ____ Magnet
10. ____ Neutral
11. ____ Peak-to-peak voltage
12. ____ Peak voltage
13. ____ Period
14. ____ Power factor
15. ____ Reactance
16. ____ RMS voltage
17. ____ Sine wave
18. ____ Single-phase voltage
19. ____ Three-phase voltage
20. ____ True power

A. Opposition in an AC circuit caused by a capacitor.
B. Opposition in an AC circuit caused by a coil or inductor.
C. The time it takes for one cycle of an AC sine wave to occur.
D. Current that changes from a positive level to a negative level periodically. Its waveform is a sine wave.
E. Root mean square is also called affective voltage. This voltage can be calculated by multiplying the peak voltage by 0.707.
F. The amount of AC voltage in one half of a sine wave that is measured from the zero point to the peak in the positive or the negative direction.
G. Power in an AC circuit that is the result of current flowing through a resistive device. It is measured in watts and does not have any current from capacitive or inductive components.
H. The opposition in an AC circuit caused by a capacitor or an inductor (coil). If voltage to a circuit is constant and reactance increases, current will decrease.
I. Voltage that is provided on two lines. For 120 volts, the lines are identified as L1 and N. For 230 volts, the lines are identified as L1 and L2.
J. The second line of a single-phase 120 volt electrical system.
K. Power caused by multiplying voltage times current when the current is caused by inductive reactance, capacitive reactance, and resistance.
L. The number of periodic cycles per unit of time.
M Voltage that is supplied as three separate lines that are identified as L1, L2, and L3. Each phase is 120° out of phase with the other.
N. A measurement of the time-phase difference between the voltage and current in an AC circuit caused when the circuit has a capacitor or inductor connected to it. It is the ratio of real power (in watts) to apparent power (in volt-amperes).
O. AC voltage that is measured between the peak of the positive half cycle and the peak of the negative half cycle.
P. An electrical switching device that is specifically designed to disconnect electrical power to a circuit or system such as an air-conditioning system or furnace.
Q. The electrical units for frequency.
R. An outline of the voltage waveform for AC voltage. The wave looks like the letter "s" that is placed on its side.
S. Any metal that can have its dipoles aligned, or any coil that produces flux lines when current flows through it.
T. Total opposition in an AC circuit caused by any combination of resistance, inductive reactance, and capacitive reactance.

TRUE OR FALSE

Place a *T* or *F* in the blank to indicate if the statement is true or false.

Score __________

1. ____ When dipoles are aligned in a material, it is a good magnet.
2. ____ Whenever current flows through a conductor, magnetic flux lines are created around the conductor.
3. ____ The polarity of the magnetic field of an electromagnet is determined by the direction of current flow through the conductor.
4. ____ Like poles of magnets attract.
5. ____ The strength of the magnetic field of a permanent magnet can be changed easily.

MULTIPLE CHOICE

Circle the letter that represents the correct answer to each question.

Score __________

1. If the size of a capacitor in an AC circuit increases from 10 μF to 20 μF, and the frequency stays the same, the capacitive reactance will:
   a. increase.
   b. decrease.
   c. remain the same.

2. When a capacitor is added to an AC circuit, the voltage waveform will _______________ the current waveform, which provides additional torque to a motor if it is connected to the circuit.
   a. lag
   b. lead
   c. remain in phase with

3. If a circuit has two large refrigeration or air-conditioning compressor motors and has a low power factor, the power factor can be raised by _______________ to make the true power equal to the apparent power, which will avoid a penalty on the electric bill.
   a. adding capacitance
   b. removing capacitance
   c. adding inductance

4. The frequency of AC voltage is defined as the:
   a. period of the voltage.
   b. number of cycles per second.
   c. peak voltage of the voltage.

5. The number of cycles per second in an AC voltage is called the _______________ of the voltage.
   a. frequency
   b. period
   c. impedance

6. The strength of the field of _______________ can be changed easily.
   a. a permanent magnet
   b. an electromagnet
   c. a dipole

7. Polarity of an electromagnet can be changed by changing the:
   a. direction of current flow through the coil of wire.
   b. amount of current flow through the coil of wire.
   c. frequency of the current flow through the coil of wire.

8. A laminated steel core is used in electromagnets that have AC voltage applied to them because laminated steel:
   a. is more economical to use than soft iron.
   b. is easily formed to any shape, which makes it more usable in complex components.
   c. allows the magnetic field to build and collapse quickly.

9. A magnetic field is created in a coil of wire when:
   a. current flows through the wire.
   b. current stops flowing through a wire.
   c. the coil has a core.

10. The left-hand rule is used to determine the:
    a. amount of magnetic flux in a coil.
    b. number of coils in an electromagnet.
    c. polarity of the magnetic field in an electromagnet.

LAB EXERCISE: MEASURE VOLTAGE IN HVAC SYSTEMS

Safety for this Lab Exercise

In this lab exercise you will be making a voltage measurement in a disconnect switch that your instructor has provided for you. The disconnect will have voltage applied to the line terminals (at the top of the disconnect switch) and when the switch is turned to the on position, it will also have voltage at the terminals at the bottom of the switch. You must be aware that the applied voltage has the potential of an electrical shock hazard and you must take care not to come into contact with the voltage anywhere the exposed electrical terminals are located. You must take care when making the voltage measurements and learn to do this task safely. This lab exercise must be fully supervised by your instructor at all times. Your instructor must initial this space indicating that he or she has discussed this lab exercise with you and is supervising your activities. _______

Tools and Materials Needed to Complete the Lab Exercise

Your instructor will provide a disconnect switch that supplies voltage to an operating air-conditioning or refrigeration system.

References to the Text

Refer to Chapter 7 in the textbook for additional information. You may need to read sections of the chapter again to help you understand the material in this exercise.

Sequence to Complete the Lab Task

Measuring Voltage in the Fused Disconnect Switch

Figure 7–1 shows a diagram that indicates where you should place your voltmeter leads when you are making a voltage test in a disconnect switch. When you begin this procedure, be sure your instructor is supervising your work. Be sure to turn the disconnect switch to the off position before you start this procedure. Notice that the terminal points at the top of the disconnect switch are identified as L1 and L2, which stands for Line 1 and Line 2. The terminals at the bottom of the disconnect are identified as T1 and T2, which stands for Terminal 1 and Terminal 2.

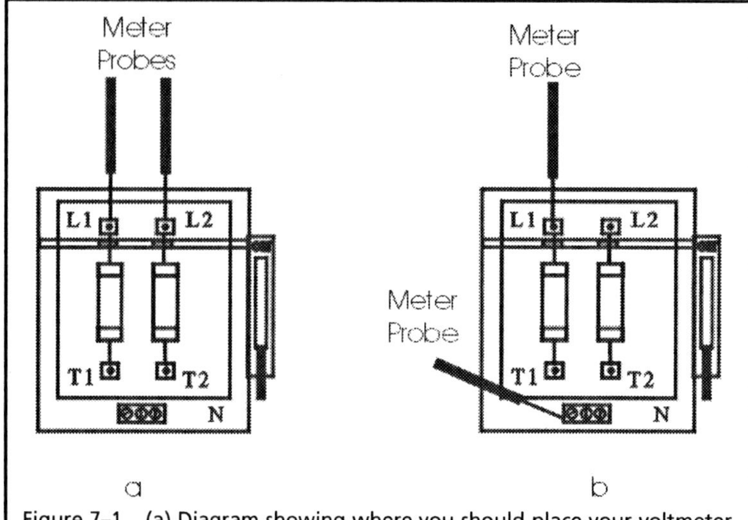

Figure 7–1 (a) Diagram showing where you should place your voltmeter terminals when measuring voltage from L1 to L2 in the disconnect switch. (b) Diagram showing where you should place your voltmeter leads when measuring voltage from L1 to N (neutral).

When the disconnect switch is turned to the off position, voltage will still be present at the top terminals, but voltage should be turned off at the fuses and at the bottom terminals of the disconnect. You will need to measure the voltage at the fuses and the terminals at the bottom of the switch to ensure voltage is turned off before you begin to work on the HVAC equipment. After you turn the switch to the off position, you can open the door and look inside it. You should notice that power comes into the switch where the wires are connected to the terminals at the top of the switch.

SAFETY NOTICE: Voltage will always be present at the terminals at the top of the disconnect switch, even when the switch is turned to the off position. The only time voltage will not be present at the top terminals in the disconnect switch is when you turn off power at the circuit breaker that supplies voltage to the disconnect switch. Always test for the presence of voltage in the disconnect and ensure it is turned off when you are working on the HVAC or refrigeration system.

1. Your instructor will provide a disconnect switch that is wired to an air conditioner or refrigeration unit. You should only work on this step when your instructor is supervising your work. Have your instructor initial in this space to indicate he or she is present during your work. _______

   In this step of the exercise, turn the disconnect to the off position and carefully open the door to the disconnect so you can view the terminals and the fuses.

   When the switch is in the open position, the switch contacts are open, and voltage is turned off to the fuses and the bottom terminals of the disconnect. This also means that voltage to the air-conditioning system or refrigeration system is also turned off. You must be aware that voltage is still present at the top terminals in the disconnect where voltage comes into the switch through the wires that are coming from the circuit breaker panel.

2. Carefully measure the voltage at the top terminals of the disconnect as shown in the Figure 7–1. Write the value of the voltage that you have measured in the spaces below.
   a. L1–L2 voltage _______________
   b. L1–N voltage _______________
   c. L2–N voltage _______________

   If the disconnect is a three-phase disconnect you can also measure these voltages.
   a. L1–L3 voltage _______________
   b. L2–L3 voltage _______________
   c. L3–N voltage _______________

3. Next, measure the voltage at the bottom of the fuses. (If the switch is open, voltage at the fuses should be zero). *Contact your instructor immediately if the voltage is not zero.*
   a. Fuse 1–Fuse 2 voltage _______________
   b. Fuse 1–N voltage _______________
   c. Fuse 2–N voltage _______________

   If the disconnect is a three-phase disconnect you can also measure these voltages.
   a. Fuse 1–Fuse 3 voltage _______________
   b. Fuse 2–Fuse 3 voltage _______________
   c. Fuse 3–N voltage _______________

4. In this step, you will measure the voltage at the bottom terminals of the disconnect. Notice that the terminals at the bottom of the disconnect switch are identified as T1 and T2, which stands for Terminal 1 and Terminal 2.
   a. T1–T2 voltage _______________
   b. T1–N voltage _______________
   c. T2–N voltage _______________

   If the disconnect is a three-phase disconnect you can also measure these voltages.
   a. T1–T3 voltage _______________
   b. T2–T3 voltage _______________
   c. T3–N voltage _______________

5. In this step, you will turn the disconnect switch back to the on position and measure the voltage at the bottom of the fuses. Now that the switch is returned to the close position, you should measure full voltage at the fuses and at the bottom terminals. *Be aware that an electrical shock hazard now exists at the top terminals, fuses, and bottom terminals. Be sure you do not come into contact with them at any time as you can receive a server electrical shock.*
   a. Fuse 1–Fuse 2 voltage _______________
   b. Fuse 1–N voltage _______________
   c. Fuse 2–N voltage _______________

If the disconnect is a three-phase disconnect you can also measure these voltages.

a. Fuse 1–Fuse 3 voltage ________________

b. Fuse 2–Fuse 3 voltage ________________

c. Fuse 3–N voltage ________________

d. T1–T2 voltage ________________

e. T1–N voltage ________________

f. T2–N voltage ________________

g. T1–T3 voltage ________________

h. T2–T3 voltage ________________

i. T3–N voltage ________________

6. Turn the disconnect switch back to the off position and have your instructor review your results. When you have completed this portion of the lab exercise, you can answer the following questions about your activities.

   a. Explain why there is still voltage present at the L1 and L2 terminals of a disconnect switch even when it is in the off position.

   ____

   ____

   b. Explain why there is no voltage at the top or bottom of the fuses when the disconnect is in the off position.

   ____

   ____

   c. Explain why there is no voltage at the bottom terminals T1 and T2 of the disconnect switch when it is in the off position.

   ____

   ____

   d. Explain why there is voltage at the fuses and at the T1 and T2 of the disconnect when it is switched to the on position.

   ____

   ____

   e. Will there be any voltage supplied to the air conditioner or refrigeration system when the disconnect switch is in the off position?

   ____

   ____

f. What position should the disconnect switch be in when you want to remove or replace electrical components in the air-conditioning or refrigeration system?

____

____

g. What position should the disconnect switch be in when you want to test fuses or remove and replace them?

____

____

Checking Out

When you have completed this lab exercise, clean up your area, return all tools and supplies to their proper place, and check out with your instructor. Your instructor will initial here to indicate you are ready to check out. _______

CHAPTER 8

Installing HVAC and Refrigeration Systems, Transformers, and Power Distribution

OBJECTIVES

At the end of this lab exercise you will be able to:

1. Correctly select the proper size wire and fuses for the installation of an HVAC.
2. Explain the operation of a control transformer.
3. Explain how to test the resistance of the primary and secondary windings of a control transformer, and determine if the transformer is good or not.

INTRODUCTION AND OVERVIEW

One of the basic tasks you will be expected to do when you are on the job is to determine the proper size of wire for an HVAC system when you are installing it. You may be able to obtain this information from the manufacturer's information, or you may have to determine the full-load amperage (FLA) rating from the system data plate and then select the wire size from the *National Electric Code*® table. One of the tasks in this lab exercise is to determine the proper wire size for a system that is being installed.

Another part of installing HVAC and refrigeration systems is to understand how transformers work. You will become more familiar with single-phase and three-phase voltages. When you are installing an HVAC or refrigeration system, it will have the voltage for the system specified on the data plate. You will need to locate a source of voltage that matches the voltage listed on the system data plate and connect your field wiring to this voltage source. As a technician you will find this voltage source in circuit breaker panels, and you will connect your field wiring to the circuit breaker and run the wire to the line-side terminals in the disconnect switch that will be mounted near the unit. It is important to understand that this voltage is supplied to the circuit breaker panel from the electrical service entrance for the residence or the business where the equipment is located.

The voltage at the service entrance comes from transformers. You will also encounter transformers in residential and commercial air-conditioning systems where they provide control voltage (usually 24 VAC) for the thermostat and system control devices. In this lab exercise you will take resistance readings of the transformer windings and also take voltage measurements at the transformer primary and secondary terminals.

TERMS

Alternator
Control transformer
Delta-connected transformer
Fuse disconnect
Primary winding
RMS voltage
Secondary winding
Step-down transformer
Step-up transformer
Transformer
Turns ratio
VA rating for transformer
Wye-connected transformer

MATCHING

Place the letter A–M for the definition from the list that matches with the terms that are numbered 1–13.

Score __________

1. ____ Alternator
2. ____ Control transformer
3. ____ Delta-connected transformer
4. ____ Fuse disconnect
5. ____ Primary winding
6. ____ RMS voltage
7. ____ Secondary winding
8. ____ Step-down transformer
9. ____ Step-up transformer
10. ____ Transformer
11. ____ Turns ratio
12. ____ VA rating for transformer
13. ____ Wye-connected transformer

A. Transformer windings that are connected in series (in the shape of the Greek letter delta).
B. An electromechanical device that produces AC voltage when its shaft is turned.
C. An electrical device with two coils (primary winding and secondary winding) that are located in close proximity. Voltage is applied to the primary winding and voltage is taken off the secondary winding.
D. Root mean square is also called affective voltage. This voltage can be calculated by multiplying the peak voltage by 0.707.
E. One of two transformer windings where voltage is taken into the transformer. Supply voltage is connected to this transformer winding.
F. The winding in the transformer where voltage comes out. Voltage is applied to the primary winding, and voltage comes out this winding.
G. A special switch that is designed to provide a disconnect for an air-conditioning system as well as provide a mounting for fuses.
H. A transformer that provides 24 volts AC secondary voltage from 120, 208, or 230 primary voltage AC.
I. The ratio of the number of turns in the primary winding as compared to the number of turns in the secondary winding.
J. The power rating of a transformer that is calculated by multiplying the voltage rating of the transformer by its amperage rating.
K. A three-phase transformer that has one end of each of its windings connected at a common point. The overall shape of the windings look like the letter "Y."
L. A transformer whose voltage in its secondary winding is lower than the voltage in its primary winding
M. A transformer whose voltage in its secondary winding is higher than the voltage in its primary winding.

TRUE OR FALSE

Place a *T* or *F* in the blank to indicate if the statement is true or false.

Score __________

1. ____ If the voltage L1–L2 is 240 V and the voltage L1–N is 120 V, the transformer is connected as a delta transformer.
2. ____ The secondary voltage is larger than the primary voltage in a step-up transformer.
3. ____ Voltage in a transformer moves from the primary winding to the secondary winding by conductance.
4. ____ The VA rating for a transformer is determined by multiplying the voltage at the secondary winding by the current flow in the secondary winding.
5. ____ The 208 volts from L2 to N for a high-leg delta transformer is usable for compressor loads that require 208 V.

MULTIPLE CHOICE

Circle the letter that represents the correct answer to each question.

Score __________

1. The voltage at the secondary terminal of a ______________ transformer will be larger than the applied voltage to the primary terminals.
   a. step-up
   b. step-down
   c. isolation

2. A transformer that is wired as a delta transformer has 480 and:
   a. 208 V.
   b. 240 V.
   c. 277 V.

3. A transformer that is wired as a wye transformer has 480 and:
   a. 110 V.
   b. 277 V.
   c. 240 V.

4. If a control transformer has a 2:1 turns ratio and it has 240 V applied to its primary winding, the voltage at the secondary winding will be:
   a. 480 V.
   b. 240 V because the secondary voltage is strictly controlled to be equal to the primary voltage.
   c. 120 V.

5. The air-conditioning or refrigeration system is connected to the ______________ of a fused disconnect.
   a. line-side terminals
   b. load-side terminals
   c. power-company terminals

LAB EXERCISE: INSTALLING AN HVAC SYSTEM

Safety for this Lab Exercise

In this lab exercise you will be making a voltage measurement on the primary and secondary terminals of a control transformer that provides 24 VAC on its secondary terminal. Your instructor will provide an operating HVAC system that has a control transformer. The voltage at the primary side of the transformer may be 230 VAC or 115 VAC, so you must be very careful to avoid any contact with the electrical terminals as you can receive a serious electrical shock. You will need to have supervision from your instructor when you make the voltage readings.

You will also be requested to make resistance tests on the primary and secondary side of a transformer. When you make the resistance tests, you will need to turn off power at the disconnect switch by turning the disconnect to the off position. You will need to take a voltage reading in the disconnect at the terminals at the bottom of the switch and ensure that voltage is turned off.

The disconnect will have voltage applied to the line terminals (at the top of the disconnect switch) even when the switch is in the off position. When the switch is turned to the on position, it will also have voltage at the terminals at the bottom of the switch. You must be aware that the applied voltage has the potential of an electrical shock hazard and you must take care not to come into contact with the voltage anywhere the exposed electrical terminals are located. You must take care when making the voltage measurements and learn to do this task safely. This lab exercise must be fully supervised by your instructor at all times. Your instructor must initial this space indicating that you are being supervised for this lab exercise. _______

Tools and Materials Needed to Complete the Lab Exercise

Your instructor will provide an operating HVAC system that has a data plate that you can read to identify the wire sizes and the FLA rating for the unit. You will also be provided with a system that has a control transformer and this system will have power applied so you can measure the voltage at the primary and secondary terminals. You will also be requested to make a resistance reading for a transformer's primary winding and secondary winding. The transformer for this exercise can either be connected in a system or it can be a part that has been removed from a system.

References to the Text

Refer to Chapter 8 in the textbook for additional information. You may need to read sections of the chapter again to help you understand the material in this exercise.

Sequence to Complete the Lab Task

Determining Correct Wire Size for Installing an HVAC or Refrigeration System

Your instructor will provide a system that has a data plate or technical data that you can read to identify the wire size for the field wiring. Field wiring is the wiring that you will add to the system to supply voltage to the system when you are installing it. For this exercise you will determine the correct wire size in two ways. First, you will use the data plate to identify the suggested wire size, and second, you will use the FLA rating to determine the wire size by using the *National Electrical Code®*. You will also be requested to identify the voltage rating for the supply voltage that you will provide for this unit. The voltage rating for wire is 300 VAC or 600 VAC. If your voltage rating for the unit is less than 300 volts, you would select wire that is rated for 300 volts. If your voltage is more than 300 volts but less than 600 volts, such as 440 volts, you would select wire, disconnect, and fuses that are all rated for 600 volts.

1. Your instructor will provide a system with a data plate or technical data that identifies the wire size. Read the data plate or technical data carefully and determine the wire size for your unit. What is the wire size for your unit? __________

2. How many wires will you need to provide voltage for your unit? __________

3. What should the color of the wire be for your unit? __________

4. Locate the full-load amperage (FLA) rating for your unit from the data plate or from the technical data. Record the FLA rating for your unit. __________

5. Use the *National Electrical Code*® table for wire sizing to determine the wire size for your unit. Record the wire size you have selected. __________

6. Check the data plate or the technical data and identify what voltage rating is for your unit. Record the voltage rating for your unit. __________

7. Based on the answer for question 6, what is the voltage rating of the wire, disconnect switch, and fuses for your unit? __________

8. Your data plate or technical data will also identify the current rating for the fuses for your unit. Identify the fuse rating for your unit and record it in this space. __________

Sequence to Complete the Lab Task

Measuring the Voltage at the Primary and Secondary Terminals of the Control Transformer to Determine if It Is Operating Correctly

Your instructor will provide an operating HVAC system that has a control transformer. You will need to have access to the primary and secondary terminals of the control transformer for this exercise so you can measure the voltage at each. If you have the correct amount of voltage at the primary terminals and at the secondary terminals, the transformer is operating correctly. If you have the proper amount of voltage at the primary terminals, but do not have voltage at the secondary terminals, the transformer is not operating correctly and it will need to be removed and replaced. You will need to have your instructor supervise this part of the activity. Have your instructor verify that you understand where to put your voltmeter terminals and how to make the voltage measurements safely. Your instructor should initial in this space to indicate you are ready to proceed with this activity. _______

> **SAFETY NOTICE:** During this exercise voltage will always be present at the primary terminals of the transformer. This voltage presents the potential for electrical shock hazard. Be sure that you do not touch any of the electrical terminals as you can receive a severe electrical shock.

1. Measure the voltage at the primary terminals of the control transformer. Record the amount of voltage you have measured. __________

2. Measure the voltage at the secondary terminals of the control transformer. Record the amount of voltage you have measured. __________

3. Refer to the voltage measurements you have made in the prior steps and identify if your transformer is operating correctly. Have your instructor check your results.

4. Explain how you have determined whether your transformer is operating correctly or not.

 ____

 ____

Sequence to Complete the Lab Task

Measuring the Resistance of the Primary and Secondary Windings of the Control Transformer to Determine if It Is Operating Correctly

In this part of the exercise you will test the resistance of the primary and secondary windings of a control transformer. Both the primary winding and the secondary winding of the transformer should have some amount of resistance. If either winding has infinite resistance, the winding has an open and will not allow current flow, and the transformer will operate correctly. You do not need to know the rating of the exact amount of resistance in the primary or secondary winding to determine it is functional. If the coils have more than 1 or 2 ohms, it indicates the winding is not shorted, and if it does not have infinite resistance you know the winding does not have an open. The exact amount of resistance is not important; rather, it is important that it does not have zero or infinite resistance, and it is considered to be functioning properly.

It is very important that you make sure the voltage to the transformer primary winding is turned off. You can do this by turning the disconnect switch to the off position. Be sure to test for voltage after you have turned the disconnect switch to the off position to ensure that the voltage is zero.

> **SAFETY NOTICE:** Voltage will always be present at the terminals at the top of the disconnect switch, even when the switch is turned to the off position. The only time voltage will not be present at the top terminals in the disconnect switch is when you turn off power at the circuit breaker that supplies voltage to the disconnect switch. Always test for the presence of voltage at the bottom terminals of the disconnect and ensure that the switch is turned to the off position and no voltage is present when you are working on the HVAC or refrigeration system.

Your instructor may prefer to give you a control transformer that is not connected in the electrical system, and you can test it on a workbench for the amount of resistance in its coils. Since the transformer is not connected to a voltage supply, you can safely make the resistance measurements.

Have your instructor initial this space to indicate that you have the disconnect switch in the off position, you do not have any voltage on the transformer windings, and it is safe to make the resistance test of the transformer windings. _______

1. Be sure that the primary windings of your transformer are isolated and disconnected so you can correctly measure the resistance in the primary winding. You only need to remove the wires from one of the primary windings and the primary winding is considered isolated. Record the amount of resistance in the primary winding. __________

2. Now you can isolate the secondary winding of the transformer by disconnecting any wires that are connected to one of the terminals of the secondary winding. You only need to isolate the wires from one of the terminals and the transformer winding will be considered isolated for the resistance test.

3. After the secondary coil is isolated, you can measure the resistance of the secondary coil of the transformer. If the coil has some amount of resistance, you know it will carry current and it will operate correctly. Record the amount of resistance you have measured in the secondary winding. __________

4. Refer to the resistance measurement you have made for your transformer and determine if the coils are functional. Remember, if the transformer coils do not have zero ohms or infinite ohms, it will produce voltage at the secondary windings when voltage is applied to the primary winding. The amount of voltage at the secondary coil will be determined by the turns ratio of the windings and the amount of voltage applied to the primary winding.

Checking Out

When you have completed this lab exercise, clean up your area, return all tools and supplies to their proper place, and check out with your instructor. Your instructor will initial here to indicate you are ready to check out. _______

CHAPTER 9

Single-Phase Open Motors

OBJECTIVES

At the end of this lab exercise you will be able to:

1. Measure the resistance of the start winding and the run winding of a single-phase open motor.
2. Identify the main parts of an AC motor and explain their functions.
3. Identify the parts of the centrifugal switch and explain their operations and functions.
4. Explain the data found on a typical single-phase motor data plate.
5. Explain the operation of a permanent split-capacitor (PSC) motor.

INTRODUCTION AND OVERVIEW

Single-phase open-type motors are used for a variety of loads in HVAC systems including evaporator fan motors, pump motors, and other types of motors. An open-type motor is a motor that has air moving over the windings for cooling, and a shaft that extends out one end. The simplest open-type motors are single-phase motors that have a start winding and run winding and a rotating shaft called an armature. The single-phase motor is also called the split-phase motor because the current in the start winding and run winding are slightly out of phase. When voltage is applied to the motor, current flows through the start winding and the run winding and causes a magnetic field to be created. The magnetic field alternates positive and then negative, which causes the armature to rotate. The current flowing in the start winding needs to be slightly out of phase with the current flowing in the run winding to get the armature to begin to rotate.

The way the motor develops a phase shift between the start winding and the run winding is that the start winding is made of very fine wire and the run winding is made of larger wire and has fewer turns than the start winding. Since the start winding is made of very fine wire, it cannot stay in the circuit and draw current for more than a few seconds that it takes to get the motor started, or it will overheat. The start winding must be electrically disconnected and de-energized once the motor gets started. The single-phase open-type motor uses a centrifugal switch that has a set of normally closed contacts and a set of flyweights to de-energize the start winding at the appropriate time. When the motor starts and comes up to speed, its armature turns at nearly 100% rpm and the rotation causes the flyweights to swing open. This action causes the centrifugal switch to move to the open position, which causes its contacts to open and de-energizes the start winding, and stops current from flowing through it. After the start winding disconnects from the circuit when the contacts of the centrifugal switch open, the motor continues to operate at full speed on only its run winding.

When the motor is ready to be stopped, voltage is turned off to the run winding and the motor shaft slows down and coasts to a stop. When the rotational speed is near zero, the flyweights of the centrifugal switch return to their starting position by spring tension, which causes the switch contacts to go back to their closed position so that the start winding can receive current when voltage is applied to the motor again on start up. You should be able to hear the centrifugal switch "click" when the motor comes up to speed, and again when the motor is de-energized and the shaft slows down to zero speed.

In this lab exercise you will have a split-phase motor that is disconnected from a power supply and placed on a bench. You will be requested to take the data from its data plate and identify the size of the motor and its rpm. In the second part of this exercise you will have a motor with its end plates removed so you can observe the centrifugal switch and flyweights. During this part of the exercise you will manually operate the flyweights and observe the centrifugal switch contacts as they open and close. You will also be requested to measure the resistance of the start winding and the run winding. In the third part of this exercise you will be provided with a third motor that is connected to a voltage supply so you are able to start the open-type motor and measure its starting current and its full-load current. When the motor is started, you will be able to listen for the "click" of the centrifugal switch, and again, when it stops.

The last part of this lab exercise will allow you to observe the parts and operation of a permanent split-capacitor (PSC) open-type motor. You will find that it does not have a centrifugal switch; rather, it uses a run capacitor to provide large current to the start winding during the time the motor shaft starts to rotate, and then the capacitor lowers the amount of current in the start winding once the motor comes up to speed.

TERMS

Ball bearing
Bushing
Capacitor-start, capacitor-run (CSCR) compressor
Capacitor-start, induction-run (CSIR) motor
Centrifugal switch
Counter EMF (CEMF)
Data plate
End plates
Four-lead motor
Frame type
Full-load amperage (FLA)
Induction motor
Locked-rotor amperage (LRA)
Motor
Multispeed motor
NEMA design
Open-type motor
Permanent split-capacitor (PSC) motor
Poles
Rotor
Run capacitor
Run winding
Service factor (SF)
Shaded-pole motor
Slip
Split-phase motor
Start capacitor
Start winding
Stator
Temperature rise

MATCHING

Place the letter A–Z for the definition from the list that matches with the terms that are numbered 1–26.

Score __________

1. ____ Centrifugal switch
2. ____ Counter EMF (CEMF)
3. ____ Data plate
4. ____ End plates
5. ____ Four-lead motor
6. ____ Frame type
7. ____ Full-load amperage (FLA)
8. ____ Induction motor
9. ____ Locked-rotor amperage (LRA)
10. ____ Motor
11. ____ Multispeed motor
12. ____ NEMA design
13. ____ Open-type motor
14. ____ Permanent split-capacitor (PSC) motor
15. ____ Poles
16. ____ Rotor

17. ____ Run capacitor
18. ____ Run winding
19. ____ Service factor
20. ____ Shaded-pole motor
21. ____ Slip
22. ____ Split-phase motor
23. ____ Start capacitor
24. ____ Start winding
25. ____ Stator
26. ____ Temperature rise

A. The amount of current a motor winding draws when power is initially applied and the rotor has not begun to rotate. This is the same amount of current the winding would draw if the rotor stopped rotating.
B. The induced voltage in an AC or DC motor that opposes the applied voltage.
C. A standardized system for motor sizes that ensures that motor mounting systems are standardized across all motor manufactures.
D. A capacitor that is connected in the start winding of a compressor or other single-phase motor. This capacitor is mounted in a plastic case and is only in the circuit for a few seconds.
E. A single-phase motor that has a run winding and a start winding.
F. When used on a motor name plate, a number that indicates how much above the name plate rating a motor can be loaded without causing serious degrading of the motor winding.
G. The difference between rotating magnetic-field speed (synchronous speed) and the rotor speed of an AC induction motor. Usually expressed as a percentage of synchronous speed.
H. The amount of temperature increase in a motor that is above the ambient temperature. The value is a rating provided for motors so that you can see how much excess temperature a motor can withstand.
I. An electromechanical device that converts electrical current to mechanical (rotational) motion at its shaft.
J. The rotating part of a motor or generator.
K. The amount of current (amperage) a motor draws when it is running at full load.
L. A motor that has a run capacitor connected in series with the start winding. The run capacitor is not disconnected from the start winding, so it stays permanently in the circuit.
M. The purpose of this switch is to open its contacts at approximately 75 to 80% of full rpm to disconnect the start winding of the motor from the source of voltage.
N. A rating standard that NEMA (National Electrical Manufacturers Association) uses to identify the different amount of starting current, toque, speeds, and other variables that are available from motors with the same horsepower rating.
O. A plate attached to a motor, transformer, or other electrical component that specifies all the electrical information (data) for the component.
P. An AC motor that operates by passing current through its wire windings (coils). When current flows through the wire coils, it creates magnetic fields.
Q. The ends of a motor that contain the bearings. The end plates are held in place by bolts that go completely through the motor from one end plate to the other.
R. A motor that has more than one speed.
S. One of two windings in a split-phase motor. The winding in a split-phase motor that has the highest resistance.
T. A capacitor that is mounted in a metal container to help dissipate heat. This type of capacitor is connected in series with the start winding of a PSC motor and remains in the circuit at all times.
U. The stationary winding of a motor.
V. A motor that is not sealed like a hermetic compressor. This type of motor is used for fans, pumps, and other general purpose applications in HVAC systems.

W. A single-phase motor with its start winding providing two leads, and its run winding providing two leads.
X. A single-phase AC induction motor that has a single winding. Each pole of the winding has a shading coil that provides the phase shift needed to allow the motor to start.
Y. The ends of magnets or magnetic fields.
Z. The main winding of a single-phase motor or compressor that has less resistance than the start winding.

TRUE OR FALSE

Place a *T* or *F* in the blank to indicate if the statement is true or false.

Score __________

1. ____ Torque is rotational force.
2. ____ Slip is the difference between the actual speed of a motor and its rated speed.
3. ____ The AC split-phase motor is called an induction motor because it uses a transformer with a capacitor for starting.
4. ____ The run winding is physically placed offset from the start winding in the stator of an AC motor to help it provide sufficient phase shift to start the rotor.
5. ____ The start winding uses larger wire than the run winding.

MULTIPLE CHOICE

Circle the letter that represents the correct answer to each question.

Score __________

1. The start capacitor is connected in series with the:
   a. run winding of a PSC motor.
   b. start winding of a CSIR motor.
   c. start winding of a shaded-pole motor.

2. The run capacitor is connected in series with the:
   a. run winding of a PSC motor.
   b. start winding of a PSC motor.
   c. start winding of the CSIR motor.

3. The shaded-pole motor:
   a. uses a capacitor and a centrifugal switch.
   b. does not have any capacitors or a centrifugal switch.
   c. uses a capacitor but does not need a centrifugal switch.

4. The permanent split-capacitor (PSC) motor uses a:
   a. run capacitor and a centrifugal switch.
   b. start capacitor and a centrifugal switch.
   c. run capacitor but does not have a centrifugal switch.

5. The capacitor-start, induction-run (CSIR) motor uses a:
   a. start capacitor that is connected in series with its start winding.
   b. start capacitor that is connected in series with its run winding.
   c. run capacitor that is connected in series with its start winding.

LAB EXERCISE: OBSERVING AND UNDERSTANDING OPEN-TYPE MOTORS

Safety for this Lab Exercise

In this lab exercise you have one or more open-type motors that are located on a workbench that you can observe the location and operation of the centrifugal switch. This motor *will not* be connected to voltage. You will be requested to measure the resistance in the start and run windings of this motor and take data from its data plate.

You instructor will provide another single-phase open motor that is connected to a power source so that you can apply voltage to the motor and turn it off. Any time you have voltage applied to the motor, you will need to be supervised by your instructor. You need to be aware of the voltage supplied to the terminals or wires of this motor so that you do not receive an electrical shock. You should also have this motor secured in a system or on a bench when it is started so that it does not move about or roll around uncontrolled. You will need to be aware of the shaft that comes out one end of the motor when it is turning so that you do not get your clothes wrapped up in it. If the shaft of the motor is connected to belts, pulleys, fans, or pumps, you will also need to be aware that you do not get your hands into the mechanical part of the system. The PSC motor uses a run capacitor to start the motor, and you must be aware that it can hold an electrical charge even after the motor is turned off, and you should not touch its terminals.

Tools and Materials Needed to Complete the Lab Exercise

Your instructor will provide a single-phase motor that is removed from its mounting and placed on a bench so that you can take it apart and measure the resistance of its start winding and run winding, observe the centrifugal switch and flyweights, and take data from its data plate. You will need a second single-phase open-type motor that can be connected to a power source so that you can observe its operation, listen to the centrifugal switch operate, and take current measurements. You will need a clamp-on ammeter for measuring the starting current and the full-load amperage (FLA). You will also need two permanent split-capacitor (PSC) motors: one that is connected to a power so you can run it, and another that is not connected to power so you can place it on a bench and inspect it.

References to the Text

Refer to Chapter 9 in the textbook for additional information. You may need to read sections of the chapter again to help you understand the material in this exercise.

Sequence to Complete the Lab Task

Taking Data from the Single-Phase Open-Type Motor's Data Plate

1. Your instructor will provide a single-phase open-type motor on a workbench so that you can take data from the data plate. Be sure that this motor is not connected to a power source and that it cannot be turned on. Find the data plate on the motor and record the following data if it is present on the data plate. Have your instructor check your information for accuracy.

    a. HP __________

    b. rpm __________

    c. Voltage __________

    d. Cycles/hertz __________

e. FLA __________
f. LRA, if provided __________
g. SF, if provided __________
h. NEMA design code __________
i. FRAME type __________
j. Temperature code __________

2. Explain why it is important to locate the data plate of a motor and record the information if you are changing out a motor.

____

____

Sequence to Complete the Lab Task

Removing the End Plate and Observing the Operation of the Centrifugal Switch and Its Flyweights

Your instructor will provide a single-phase open-type motor that has a centrifugal switch for energizing and de-energizing the start winding into the circuit.

1. Carefully remove the end plate from the motor on the end opposite the shaft.

2. When you have the end plate removed, you should be able to locate the flyweights and the normally closed contacts of the centrifugal switch.

3. Reach into the motor and move the flyweights outward to the position they would be at when the motor reaches full speed. Observe that the normally closed contacts of the starting switch will move to the open position when the flyweights move outward.

4. Allow the flyweights to return to their original position that they assume when the motor shaft stops rotating. Notice the contacts of the centrifugal switch return to their closed position.

5. Connect an ohmmeter to the two leads of the start winding so you can observe the centrifugal switch in the closed position. The resistance should be approximately 10 to 20 ohms. Measure the resistance of the start winding and record it in this space. __________

6. Move the centrifugal switch flyweights to the open position so that the centrifugal switch contacts open. Measure and record the resistance of the winding when the switch contacts are in the open position. __________

7. Allow the flyweights to return to their normal position and record the start winding resistance again. __________

8. Carefully place the end plate back onto the motor so it will be ready for the next person doing this exercise.

9. What can you say about the amount of resistance the start winding has when the centrifugal switch is closed and when it is open?

____

____

10. What do theses resistance readings indicate in regard to the start winding being in or out of the circuit as the centrifugal switch opens and closes?

____

____

Sequence to Complete the Lab Task

Measuring the Resistance of the Run Winding and Start Winding

1. In this step, you will locate the wires or terminals of the motor that are connected to the run winding. Measure the resistance of the run winding and record the amount of resistance. __________
2. Locate the wires connected to the start winding and measure the resistance of the start winding and record it here. __________
3. Compare the resistance in the start and run windings. Which winding has the most resistance? __________
4. If you are troubleshooting a split-phase motor and you measure the resistance of the run winding and the start winding, what resistance values would you expect to measure for the run winding and the start winding if the motor windings are good?
   a. Run winding resistance should be about __________ ohms.
   b. Start winding resistance should be about __________ ohms.
5. Explain how you could use the resistance measurement to troubleshoot a split-phase motor to determine if the centrifugal switch and the start and run windings are good.

____

____

Sequence to Complete the Lab Task

Starting the Single-Phase Open-Type Motor and Listening for the Operation of the Centrifugal Switch

1. This portion of this lab exercise should be fully supervised by your instructor. Have your instructor initial here when you are ready to complete this part of the exercise. _______
2. Make sure the motor you are going to apply power to is mounted securely and that there are no obstructions with its shaft. Turn the power switch to the on position and the motor should start. Listen closely for the "click" of the centrifugal switch a few seconds after the motor starts. Indicate in this space what the operation of the centrifugal switch sounds like when it is working correctly.

____

____

3. Turn power to the motor off and let the motor coast to a stop. Listen closely for the centrifugal switch to close again when the motor shaft coasts to a stop. Explain when the centrifugal switch will return to the closed position.

____

____

4. Explain how you can tell if the centrifugal switch is operating correctly by listening to it.

____

____

Sequence to Complete the Lab Task

Starting the Single-Phase Open-Type Motor and Measuring the Starting Current and the Full-Load Current (FLA) with a Clamp-on Ammeter

1. This portion of the lab exercise should be fully supervised by your instructor. Have your instructor initial here when you are ready to complete this part of the exercise. _______

2. In this part of the lab exercise you are going to use a clamp-on ammeter to measure the current in the start winding and the run winding. Your instructor will provide a motor that you can turn on and off to measure the current. Be sure to have the power turned off to the motor, and place the clamp-on ammeter around one of the wires going to the start winding. When the clamp-on ammeter is in place around one of the wires for the start winding, turn power on and record the current flowing in the start winding and record it here. __________

3. Turn the power to the motor off and let the motor stop. This time when you start the motor, allow it to run for two minutes. Observe the current in the start winding and report what you have observed.

____

____

4. Turn the power to the motor off again and let the motor stop. Place the clamp-on ammeter around the winding to the run winding. Start the motor again and allow it to run for two minutes. Observe the current in the run winding and report what you have observed.

____

____

5. Turn the power to the motor off again and let the motor stop. Place the clamp-on ammeter around the L1 power line that supplies voltage to both the start winding and the run winding. Start the motor again and allow it to run for two minutes. Record the current when the motor first starts. This is called the locked-rotor amperage (LRA). __________

6. Record the motor current after the motor has come up to speed and the start winding has been disconnected by the centrifugal switch. This current is called the full-load amperage (FLA). __________

7. Compare the LRA to the FLA and explain which one is larger, and why.

____

____

8. Explain what the problem would be if the motor continued to pull LRA values after it first starts and did not return to FLA values.

___

___

9. Explain how you could use the clamp-on ammeter for troubleshooting a single-phase open-type motor.

___

___

Sequence to Complete the Lab Task

Starting the Permanent Split-Capacitor (PSC) Motor and Measuring the Starting Current and the Full-Load Current (FLA) with a Clamp-on Ammeter

1. Locate the PSC motor that is provided on a bench for you to inspect. You should notice that the PSC motor does not have a centrifugal switch. The motor has three terminals or wires that are identified as *R* for run winding, *S* for start winding, and *C* for the common terminals where the run and start windings are connected. You can refer to the diagrams in Chapter 9 of the textbook. Use an ohmmeter and measure the resistance in the run winding (terminal R to C). ________

2. Use an ohmmeter to measure the resistance in the start winding (terminal S to C).

   ____

3. How can you use this information about the resistance readings to determine which motor terminal is the run winding and which is the start winding?

   ___

   ___

4. During this next step of the procedure, you will need to be supervised by your instructor. Have your instructor initial in this space to indicate you are being supervised. ______

5. Locate the PSC motor that is connected to power and place the clamp-on ammeter around the wire connected to the run winding. Apply power to the PSC motor and measure the current to the run winding and record it here. ________

6. Turn power off and place the clamp-on ammeter around the wire connected to the start winding. Turn power on to the motor and observe the current level when the motor first starts, and then again when it has been running for one minute.

   a. Record the current when the motor first starts. ________

   b. Record the current when the motor has run for one minute. ________

7. What can you say about the two levels of current you measured in step 6?

   ___

   ___

8. The reason the current is lower once the motor comes up to full speed is that the run capacitor provides capacitive reactance to the current in the start winding circuit, which acts like a resistance that lowers the current. The capacitor is called a run capacitor because it remains connected to the start winding all the time the motor is running. Locate the run capacitor on your PSC motor and record the following.
   a. Capacitance of the capacitor in microfarads __________
   b. The voltage rating of the capacitor __________
   c. What is the material that the outside of the capacitor is made of? __________
   d. Explain why this material allows heat to dissipate easily.

   ____

   ____

9. If you have a multimeter that can measure capacitance, disconnect the capacitor from the motor and measure its capacitance and record its value here. __________

10. In this step, you are going to try to start the PSC motor with one of the leads to the capacitor disconnected. Have your instructor disconnect the lead from the capacitor that goes to the start winding of the motor. Be sure to place this wire in a place where it will not come into contact with any metal on the motor or unit, or with your skin, as it will have full voltage at its end. When you try to start the motor, it will not start because the disconnected wire causes an open in the start winding, and it will cause excessive current in the run winding. For this reason, you should only apply power for a few seconds so that you can measure the current and then turn power to the motor off. Place the clamp-on ammeter around the L1 wire that feeds power to the motor and apply power to the motor for two to three seconds as you make the current measurement. Immediately turn off all power to the motor so that you do not overheat it. What is the amount of current that the motor draws when it tries to start?

    ____

11. What kind of noise does the motor make when it tries to start and cannot run? __________

12. Explain why the PSC motor will not have any problems with a starting switch like the split-phase motor.

    ____

    ____

13. Explain why the most likely cause of the PSC motor not starting will be a bad or faulty run capacitor or start winding.

    ____

    ____

Checking Out

When you have completed this lab exercise, clean up your area, return all tools and supplies to their proper place, and check out with your instructor. Your instructor will initial here to indicate you are ready to check out. _______

CHAPTER 10

Single-Phase Hermetic Compressors

OBJECTIVES

At the end of this lab exercise you will be able to:

1. Measure the resistance of the start winding and the run winding of a single-phase hermetic compressor.
2. Identify the parts of the current relay and explain its operation and function.
3. Explain the data found on a typical single-phase hermetic compressor data plate.
4. Explain the operation of a permanent split-capacitor (PSC) compressor.
5. Explain the operation of a potential relay and the start capacitor.
6. Explain the operation of a hermetic compressor overload.

INTRODUCTION AND OVERVIEW

A single-phase hermetic compressor is a single-phase motor that is sealed inside a container and the extra refrigerant that is returned to the compressor is used to cool its windings. The hermetic compressor is similar to the single-phase open-type motor in that it has a start winding, a run winding, and a rotating armature, but in the hermetic compressor, the armature shaft turns a compressor that pumps refrigerant through the HVAC system. The single-phase hermetic compressor is also called the split-phase compressor because the current in the start winding and run winding are slightly out of phase. When voltage is applied to the motor, current flows through the start winding and the run winding and causes a magnetic field to be created. The magnetic field alternates positive and then negative, which causes the armature to rotate. The current flowing in the start winding needs to be slightly out of phase with the current flowing in the run winding to get the armature to begin to rotate.

The way the motor develops a phase shift between the start winding and the run winding is that the start winding is made of very fine wire and the run winding is made of larger wire and has fewer turns than the start winding. Since the start winding is made of very fine wire, it cannot stay in the circuit and draw current for more than the few seconds that it takes to get the motor started, or it will overheat. The start winding must be electrically disconnected and de-energized once the motor gets started. Unlike the single-phase open-type motor that uses a centrifugal switch to disconnect the start winding, the hermetic compressor needs a current relay or potential relay to connect the start winding when the motor is starting, and then disconnect it several seconds later once the motor comes up to speed. The hermetic compressor, like the split-phase open-type motor, runs on only the run winding after the start winding is disconnected from the circuit and the motor continues to operate at full speed.

A small hermetic compressor may be able to operate as a split-phase motor and have sufficient torque to start, but larger split-phase compressors that are between .5 and 2 HP may need one or more capacitors to add starting torque. Starting torque is the extra rotational force that is needed to get the piston to start moving when the motor first starts. Since the refrigerant has pressure that is exerted on the pumping part of the compressor, it resists the compressor motor shaft from turning when it is first starting. If a start capacitor or run capacitor is added to the motor, it will have additional starting torque. If a start capacitor and a run capacitor are both added, the single-phase compressor will be able to develop more torque.

In this lab exercise you will have a split-phase hermetic compressor motor that is disconnected from a power supply and placed on a bench. You will be requested to take the data from its data plate and identify the size of the motor and its current and voltage ratings. In the second part of this exercise you will ensure the compressor motor is disconnected from power so you can measure the resistance in its start and run windings.

In the next section of this lab exercise, you will be provided a single-phase compressor that uses a current relay to control the start winding. You will also be provided a single-phase compressor that has a start capacitor and a current relay, as well as a single-phase compressor that uses a start capacitor, run capacitor, and a potential relay to develop more torque when it is started. The last type of single-phase compressor motor you will learn about will be connected as a permanent split-capacitor (PSC) type motor with only a run capacitor connected to its start winding.

In each section of the lab exercise, you will be provided a diagram of the way the motor is connected that shows all the components and the windings. You will also be requested to measure the resistance of the start winding, the run windings, and any other component. You will also be requested to use a clamp-on ammeter to test the current in the run and start windings of each type of compressor. The resistance tests and the current measurements will be used to troubleshoot the compressors you encounter in HVAC service calls.

The last part of this exercise is to test the overload for the hermetic compressor. The overload is a normally closed set of contacts that is wired in series with terminal C of the compressor, and it can sense overcurrent as well as excessive temperature in the winding.

TERMS

Bimetal overload
Capacitor-start, capacitor-run (CSCR) compressor
Clamp-on ammeter
Common terminal C
Compressor shell
Compressor terminals
Current relay
Full-load amperage (FLA)
Hard-start kit
Hot wire relay
Induction
Locked-rotor amperage (LRA)
Overload
Permanent split-capacitor (PSC) compressor
Potential relay
Run capacitor
Run terminal R
Run winding
Single-phase compressor
Solid state starting relay
Split-phase compressor
Split-phase hermetic compressor
Start capacitor
Start terminal S
Start winding
Thermal overload
Troubleshooting the compressor

MATCHING

Place the letter A–X for the definition from the list that matches with the terms that are numbered 1–24.

Score __________

1. ____ Bimetal overload
2. ____ Capacitor-start, capacitor-run (CSCR) compressor
3. ____ Clamp-on ammeter
4. ____ Common terminal C
5. ____ Compressor shell
6. ____ Compressor terminals
7. ____ Current relay
8. ____ Full-load amperage (FLA)
9. ____ Hard-start kit
10. ____ Hot wire relay
11. ____ Induction
12. ____ Locked-rotor amperage (LRA)
13. ____ Overload
14. ____ Potential relay
15. ____ Permanent split-capacitor (PSC) compressor
16. ____ Run capacitor
17. ____ Run terminal R
18. ____ Run winding
19. ____ Single-phase compressor
20. ____ Solid-state starting relay
21. ____ Start capacitor
22. ____ Start terminal S
23. ____ Start winding
24. ____ Thermal overload

A. The amount of current (amperage) a motor draws when it is running at full load. This is also the rating used to determine the proper fuse, circuit breaker, or motor starter for a motor.

B. An ammeter that has a set of "jaws" that open to allow them to wrap around a wire. The jaws close around the wire and function as a transformer coil to sense the amount of current flowing through the wire.

C. An electrical overload device that is composed of two dissimilar metals. One part converts electrical current into heat, and the second part of the overload senses the increase in temperature and moves.

D. A device consisting of a capacitor and solid state control that is used to help start a single-phase compressor motor.

E. A single-phase compressor motor that has a start capacitor and a run capacitor. A run capacitor is permanently connected in series with the start winding, and the start capacitor is disconnected from the circuit by a potential relay once the compressor comes up to speed. The start capacitor and run capacitor creates the most torque in a single-phase compressor.

F. A device used to protect motors that sense current flowing to the motor. When the current flows through the overload, it produces heat.

G. The main winding of a single-phase motor or compressor that has less resistance than the start winding.

H. A special starting relay used to energize and de-energize the start winding on a compressor. Its contacts are normally closed, and provide a connection for the start capacitor to the start winding.

I. The lead that is connected to the start winding of a single-phase compressor motor.

J. The point in single-phase motor windings where the start winding and run winding are connected.

K. The amount of current a motor winding draws when power is initially applied and the rotor has not begun to rotate. This is the same amount of current the winding would draw if the rotor stopped rotating.
L. The electrical terminals that are mounted on the side of the compressor shell.
M. A capacitor that is mounted in a plastic case and is connected in the start winding of a compressor or other single-phase motor. The start capacitor is only in the circuit for a few seconds.
N. A capacitor that is mounted in a metal container to help dissipate heat. It is connected in series with the start winding of a PSC motor and remains in the circuit at all times.
O. A special relay that has a single set of normally closed contacts, and is used to start a compressor by controlling the start winding of a sealed compressor motor.
P. The terminal connected to the run winding or main winding of the motor or compressor.
Q. The metal container that houses the compressor motor and pump. It is sealed so that it does not leak refrigerant.
R. A relay that is designed to use solid state components to provide starting current to the start winding of a single-phase motor, much like a current relay.
S. A part of the motor starter that consists of a heater element and a set of normally closed contacts.
T. One of two windings in a split-phase motor. This winding in a split-phase motor has the highest resistance.
U. Creating current in a coil of wire by passing it through a magnetic field.
V. A compressor motor that has a run capacitor connected in series with the start winding. The run capacitor is not disconnected from the start winding, so it stays permanently in the circuit.
W. A special motor starting relay that was used extensively to start single-phase hermetic compressors before the advent of electronic controls. It consists of a heating element and contacts that change position from normally closed to open when current flows through the heating element.
X. A compressor motor that runs on single-phase voltage. The compressor motor has a run winding and start winding that are connected at a point called common C.

TRUE OR FALSE

Place a *T* or *F* in the blank to indicate if the statement is true or false.

Score __________

1. ____ The run winding or the start winding of a single-phase compressor will draw current if it has an open.

2. ____ The potential relay contacts are normally closed and are opened when the back EMF (CEMF) in the motor windings becomes large enough to energize the relay coil when the motor reaches 75% of full speed.

3. ____ The run capacitor in the PSC compressor remains in the circuit at all times and can help regulate the speed of the compressor.

4. ____ The contacts of the current relay are normally closed.

5. ____ The thermal overload on the hermetic compressor can be opened by either excessive current draw or overheating of the compressor.

MULTIPLE CHOICE

Circle the letter that represents the correct answer to each question.

Score __________

1. The current relay has a coil with:
   a. low resistance and normally closed contacts.
   b. high resistance and normally closed contacts.
   c. low resistance and normally open contacts.
   d. high resistance and normally open contacts.

2. The potential relay has a coil with:
   a. low resistance and normally closed contacts.
   b. high resistance and normally closed contacts.
   c. low resistance and normally open contacts.
   d. high resistance and normally open contacts.

3. The split-phase hermetic compressor motor uses a:
   a. current relay that has its coil connected in series with the start winding and its contacts connected in series with the run winding.
   b. current relay that has its coil connected in series with the run winding and its contacts connected in series with the start winding.
   c. potential relay that has its contacts connected in series with the start windings.

4. The CSIR split-phase hermetic compressor motor uses a:
   a. current relay that has its coil connected in series with the run winding and its contacts connected in series with the start winding.
   b. current relay that has its coil connected in series with the start winding and its contacts connected in series with the run winding.
   c. potential relay that has its contacts connected in series with the start windings.

5. The CSCR split-phase hermetic compressor motor uses a:
   a. potential relay that has a run capacitor connected in series with the run winding, and a start capacitor connected in series with the potential relay normally open contacts and start winding.
   b. current relay that has its coil connected in series with the start winding and its contacts connected in series with the run winding.
   c. potential relay that has its contacts connected in series with the start capacitor and start winding, and a run capacitor connected between the run and start windings.

LAB EXERCISE: OBSERVING AND UNDERSTANDING HERMETIC-TYPE COMPRESSORS

Safety for this Lab Exercise

In this lab exercise you have one or more hermetic-type compressors that are located on a work-bench that *will not* be connected to voltage. You will be requested to measure the resistance in the start and run windings of these motors and take data from their data plates.

Your instructor will provide additional single-phase hermetic compressors that are connected to a power source so that you can apply voltage to the compressors and observe their operation. Any time you have voltage applied to the compressors, you will need to be supervised by your instructor.

You need to be aware of the voltage supplied to the terminals or components of these compressors so that you do not receive an electrical shock. The capacitor-start compressor, the capacitor-start, capacitor-run compressor, and the PSC compressors all use capacitor to start the motor, and you must be aware that the capacitors can hold an "electrical charge" even after the compressor is turned off, and you should not touch its terminals.

Tools and Materials Needed to Complete the Lab Exercise

Your instructor will provide a single-phase hermetic compressor that is removed from an HVAC or refrigeration system and place it on a bench so that you can measure the resistance of its start and run windings, check out the components used to help it start, and take data from its data plate. You will need additional hermetic compressors that are wired as capacitor-start compressor, capacitor-start, capacitor-run compressor, and PSC compressor. These compressors need to be in an operational system so that you can connect power and run them. When the compressors are running, you will be requested to take current measurements. You will need a clamp-on ammeter for measuring the starting current and the full-load amperage (FLA).

References to the Text

Refer to Chapter 10 in the textbook for additional information. You may need to read sections of the chapter again to help you understand the material in this exercise.

Sequence to Complete the Lab Task

Taking Data from the Single-Phase Hermetic Compressor's Data Plate

1. Your instructor will provide a single-phase hermetic compressor on a workbench so that you can take data from the data plate. Be sure that this compressor is not connected to a power source and that it cannot be turned on. Find the data plate on the compressor and record the following data if it is present on the data plate. Have your instructor check your information for accuracy.

   a. HP __________

   b. Voltage __________

   c. Cycles/hertz __________

   d. FLA __________

   e. LRA, if provided __________

2. Explain why it is important to locate the data plate of a compressor and record the information if you are changing it out.

   ____

   ____

Sequence to Complete the Lab Task

Measuring the Resistance of the Run Winding and Start Winding of a Hermetic Compressor

Use the diagrams shown in Figures 10–1 and 10–2 for the resistance measurements in this section. You should notice from these diagrams that the left end of the run winding is connected to the R terminal on the compressor terminal block, and the right end of the run winding is connected to the right end of the start winding, and this point is identified as the common terminal or common point. The right end of the start winding is connected to the S terminal of the compressor terminal block. This means that you can test for the resistance of the run winding by putting one ohmmeter lead on terminal R and the other on terminal C. You can measure the resistance of the start winding by putting one terminal of the ohmmeter on terminal S and the other on terminal C. The reason terminal C is called the common point is that it is used in common by both the run winding and the start winding.

The run winding is made with fewer coils of larger wire than the start winding. You should notice the resistance of the run winding is less than the resistance of the start winding.

1. Your instructor will provide a hermetic compressor that is not connected to power that you will use to measure and record the resistance of the run winding and the start winding. Record your measurements here.

    a. Run winding resistance __________

    b. Start winding resistance __________

2. Which winding has the most resistance, and why?

    ____

    ____

3. Explain how you could use the resistance measurement to troubleshoot a split-phase compressor to determine if the start and run windings are good.

    ____

    ____

4. If you are troubleshooting a split-phase compressor motor and you measure the resistance of the run winding and the start winding, what resistance values would you expect to measure for the run winding and the start winding if the motor windings are good?

    a. Run winding resistance should be about _______ ohms.

    b. Start winding resistance should be about _______ ohms.

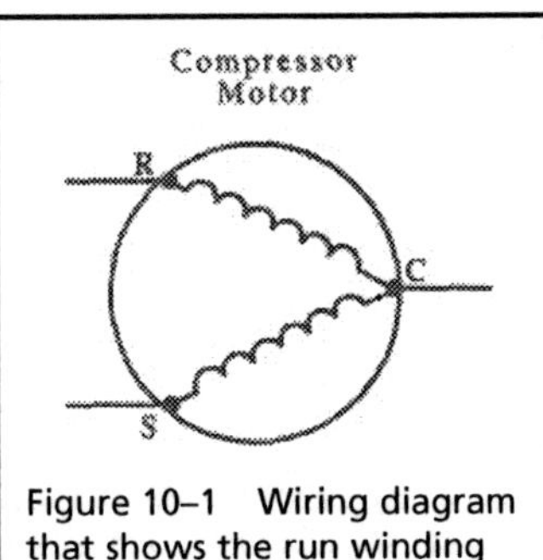

Figure 10–1 Wiring diagram that shows the run winding and start winding of a hermetic compressor.

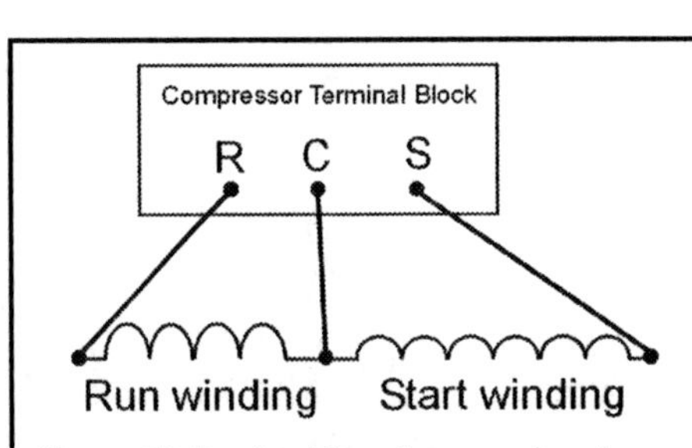

Figure 10–2 A wiring diagram showing the run winding, start winding, and common terminals for a hermetic compressor.

Sequence to Complete the Lab Task

Identifying the Run Winding and Start Winding of a Hermetic Compressor from Resistance Readings

Sometimes you will encounter a compressor motor where it is difficult to determine which is the R, S, or C terminal. One way to positively determine the terminals for each winding is to measure their resistance between the terminals and draw the values on a piece of paper. In Figure 10–3, you can see that the values between the three unknown terminals are as follows: between terminals 1 and 3, you have 14 ohms; between terminals 1 and 2, you have 4 ohms; between terminals 2 and 3, you have 10 ohms.

You can use your knowledge of the comparison of the resistance of the run and start windings to identify the R, S, and C terminals. Since you know the lowest value of resistance will be the run winding, you know the run winding has to be between terminals 1 and 2. At this point, you know the run winding is found between terminals R and C, but you do not know whether terminal 1 is the R terminal or the C terminal. You know the highest value of resistance is the combination of the run and start windings that would be measured between the R terminal and the S terminal, which we know are terminals 1–3. Since we know that terminals 1and 2 are the run windings, it means that terminal 3 is the start winding terminal. The middle resistance value (10 ohms) must be the start winding, and it is measured between terminals 2 and 3. This means that the start winding must be terminal 3, and the common terminal is terminal 2. You can use this process of elimination to determine the start, run, and common terminals of any single-phase compressor motor.

1. In the space below, identify the start, run, and common terminals of this compressor if the following resistance is measured between the following terminals.
   a. Resistance between terminals 1–2 is 18 ohms. __________
   b. Resistance between terminals 2–3 is 6 ohms. __________
   c. Resistance between terminals 1–3 is 12 ohms. __________
   d. Terminal 1 is the __________ terminal.
   e. Terminal 2 is the __________ terminal.
   f. Terminal 3 is the __________ terminal.

Sequence to Complete the Lab Task

Starting the Permanent Split-Capacitor (PSC) Compressor and Measuring the Starting Current and the Full-Load Current (FLA) with a Clamp-on Ammeter

1. This section of this lab exercise should be fully supervised by your instructor. Have your instructor initial here when you are ready to complete this part of the exercise. _______

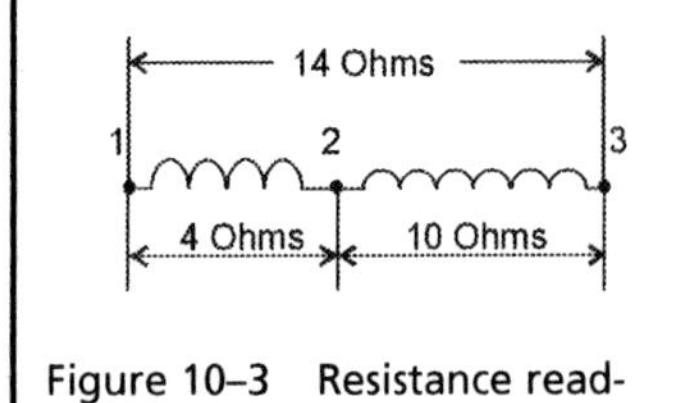

Figure 10–3 Resistance readings from terminals 1, 2, and 3 that are unknown.

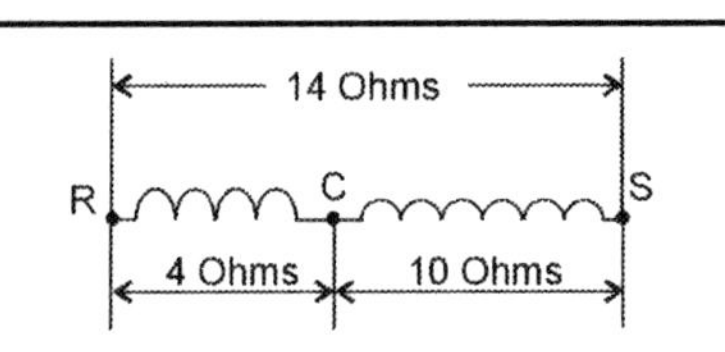

Figure 10–4 Resistance measurement used to identify R, S, and C terminals of a hermetic compressor.

Your instructor will provide a compressor wired as a PSC compressor. The PSC compressor provides more starting torque than a split-phase compressor that does not use any capacitors. Figure 10–5 shows the electrical diagram of the PSC compressor. You should notice that the PSC compressor has three terminals that are identified as *R* for run winding, *S* for start winding, and *C* for the common terminals, where the run and start windings are connected. From the diagram you can see that the run capacitor is connected in series with the start winding.

2. Place the clamp-on ammeter around the wire connected to the run winding. Apply power to the PSC compressor and measure the current to the run winding and record it here.

   ____

3. Turn power off and place the clamp-on ammeter around the wire connected to the start winding. Turn power on to the motor and observe the current level when the motor first starts, and then again when it has been running for one minute.

   a. Record the current when the motor first starts. __________

   b. Record the current when the motor has run for one minute. __________

4. What can you say about the two levels of current you measured in step 3?

   ____

   ____

5. The reason the current is lower once the motor comes up to full speed is that the run capacitor provides capacitive reactance to the current in the start winding circuit, which acts like a resistance that lowers the current. The capacitor is called a run capacitor because it remains connected to the start winding all the time the motor is running. Locate the run capacitor on your PSC motor and record the following.

   a. Capacitance of the capacitor in microfarads __________

   b. The voltage rating of the capacitor __________

   c. What is the material that the outside of the run capacitor is made of? __________

   d. Explain why this material allows heat to dissipate easily.

   ____

   ____

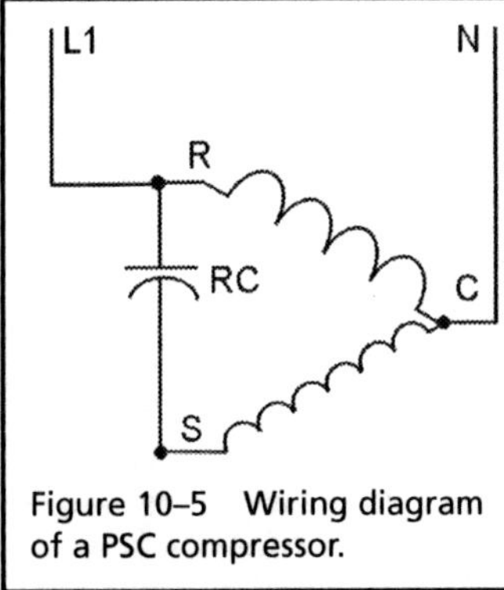

Figure 10–5 Wiring diagram of a PSC compressor.

e. If you have a multimeter that can measure capacitance, disconnect the capacitor from the motor and measure its capacitance and record its value here. __________

Sequence to Complete the Lab Task

Starting the Single-Phase Hermetic Compressor and Observing the Operation of the Current Relay

1. This section of this lab exercise should be fully supervised by your instructor. Have your instructor initial here when you are ready to complete this part of the exercise. _______

   In this part of the lab exercise you are going to use a clamp-on ammeter to measure the current in the start winding and in the run winding of a hermetic compressor that has its start winding controlled by a current relay. Figure 10–6 shows a wiring diagram of this motor. The current relay has a set of normally open contacts that are wired in series with the start winding, and a coil that is wired in series with the run winding. When voltage is applied to the motor, a large amount of locked-rotor amperage (LRA) will begin to flow in the run winding, which will also cause the same current to flow through the current relay coil. This current is strong enough to cause the normally open contacts to close, and allows voltage to be supplied to the start winding. When voltage is applied to the start winding, it begins to draw current and the motor shaft will begin to turn. When the motor shaft comes up to full speed, the current in the run winding will return from the larger LRA to a smaller amount of current called full-load amperage (FLA). When the current in the run winding reduces to the FLA level, the spring in the current relay pulls its contacts back to their open position, which interrupts voltage to the start winding and effectively disconnects the start winding so it does not overheat.

2. Your instructor will provide a hermetic compressor with a current relay that is connected to a power source so you can turn the power on and off and measure the current that flows through the run and start windings. Be sure to have the power turned off to the motor and place the clamp-on ammeter around one of the wires going to the start winding. When the clamp-on ammeter is in place around the wire for the start winding, turn power on and record the current flowing in the start winding. Record it here when the motor starts, and again when the motor has runs for a few seconds and comes up to speed.

   a. What is the current when voltage is first applied? __________

   b. What is the current when the motor comes up to speed? __________

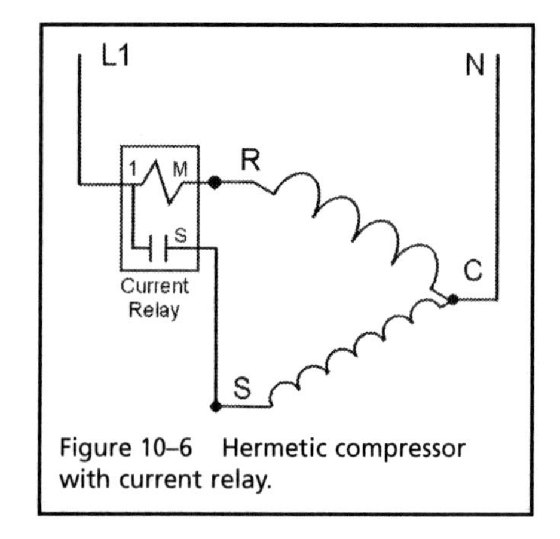

Figure 10–6 Hermetic compressor with current relay.

3. Turn the power to the motor off and let the motor stop. Place the clamp-on ammeter around the winding to the run winding. Start the motor again and allow it to run for 20 seconds and observe the current in the run winding. Report what you have observed.

   a. What is the current level when the motor first starts? __________

   b. What is the current level when the motor has run for 20 seconds? __________

4. Turn the power to the motor off again and let the motor stop. Place the clamp-on ammeter around the L1 power line that supplies voltage to both the start winding and the run winding. Start the motor again and allow it to run for two minutes. Record the current when the motor first starts. This is called the locked-rotor amperage (LRA). __________

5. Record the motor current after the motor has come up to speed and the start winding has been disconnected by the current relay. This current is called the full-load amperage (FLA). __________

6. Compare the LRA to the FLA and explain which one is larger, and why.

   ____

   ____

7. Explain what the problem would be if the motor continued to pull LRA values after it first starts and did not return to FLA values.

   ____

   ____

8. Explain how you could use the clamp-on ammeter for troubleshooting a single-phase compressor.

   ____

   ____

Sequence to Complete the Lab Task

Starting the Capacitor-Start Induction-Run Hermetic Compressor and Observing the Operation of the Current Relay

1. This section of the lab exercise should be fully supervised by your instructor. Have your instructor initial here when you are ready to complete this part of the exercise. _______

   In this part of the lab exercise you are going to use a clamp-on ammeter to measure the current in the start winding and in the run winding of a capacitor-start induction-run hermetic compressor that has its start winding controlled by a current relay. Figure 10–7 shows a wiring diagram of this motor. The current relay has a set of normally open contacts that are wired in series with the start capacitor and the start winding. The coil of the current relay is wired in series with the run winding. When voltage is applied to the motor, a large amount of locked-rotor amperage (LRA) will begin to flow in the run winding, which will also cause the same current to flow through the current relay coil. This current is strong enough to cause the normally open contacts to close, and allows voltage to be supplied to the start winding and the start capacitor. When voltage is applied to the start winding and start capacitor, a phase shift develops that helps create additional starting torque and helps the motor shaft turn. When the motor shaft comes up to full speed, the current in the run winding will return from the larger amount of LRA to a smaller amount of current called full-load amperage (FLA). When the current in the run winding reduces to the FLA level, the spring in the current relay pulls its contacts back to their open position, which interrupts voltage to the start winding and start capacitor and effectively disconnects the start winding so it does not overheat.

2. Make sure power is turned off to the compressor and check the rating on the start capacitor. Record the following data.
   a. Microfarad rating __________
   b. Voltage rating ____________

3. Your instructor will provide a hermetic compressor with a current relay that is connected to a power source so you can turn on and off the power and measure the current that flows through the run and start windings. Be sure to have the power turned off to the motor and place the clamp-on ammeter around one of the wires going to the start winding. When the clamp-on ammeter is in place around the wire for the start winding, turn power on and record the current flowing in the start winding and record it here when the motor starts, and again when the motor has runs for a few seconds and comes up to speed.
   a. What is the current when voltage is first applied? __________
   b. What is the current when the motor comes up to speed? __________

4. Turn the power to the motor off and let the motor stop. This time place the clamp-on ammeter around the winding to the run winding. Start the motor again. Allow it to run for 20 seconds, and report what you have observed.
   a. What is the current level when the motor first starts? __________
   b. What is the current level when the motor has run for 20 seconds? __________

5. Turn the power to the motor off and let the motor stop. Place the clamp-on ammeter around the L1 power line that supplies voltage to both the start winding and the run winding. Start the motor and allow it to run for two minutes. Record the current when the motor first starts. This is called the locked-rotor amperage (LRA). __________

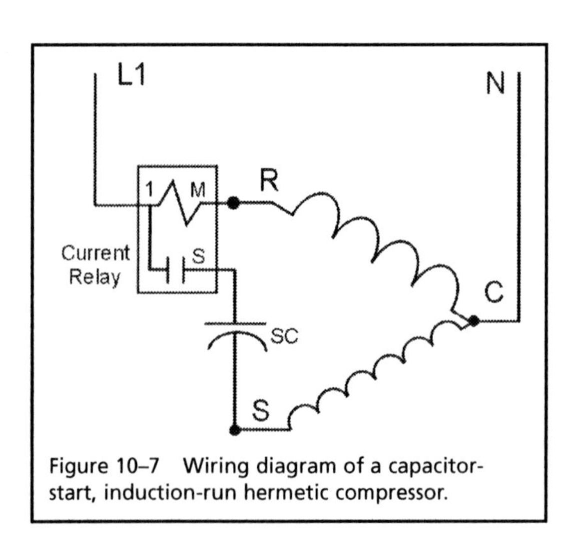

Figure 10–7 Wiring diagram of a capacitor-start, induction-run hermetic compressor.

6. Record the motor current after the motor has come up to speed and the start winding has been disconnected by the current relay. This current is called the full-load amperage (FLA). __________

7. Compare the LRA to the FLA and explain which one is larger, and why.

____

____

8. Explain what the problem would be if the motor continued to pull LRA values after it first starts and does not return to FLA values.

____

____

9. Explain how you could use the clamp-on ammeter for troubleshooting a single-phase compressor.

____

____

Sequence to Complete the Lab Task

Starting the Capacitor-Start, Induction-Run Hermetic Compressor and Observing the Operation of the Current Relay

In this part of the lab exercise you will inspect a current relay and identify coil and contacts. You will also use an ohmmeter to measure the resistance in the coil and the contacts.

1. Figure 10–8 shows a picture of a current relay. Place an arrow on the picture of the current relay in Figure 10–8 and identify the coil and the location of the contacts. Have your instructor check your work for accuracy.

2. Your instructor will provide a current relay that is not connected to a compressor or to any power source, so you can visually inspect its coil and contacts. Use an ohmmeter to measure the resistance of the coil and record it in this space. __________

3. Use an ohmmeter to measure the resistance of the contacts of the current relay and indicate if the contacts are normally open or normally closed. Be aware that the contacts of the current relay are sensitive to location. Be sure you have the relay positioned so that the arrow on the side of the relay is pointing up. Are the contacts normally open or normally closed?

____

Figure 10–8 A picture of a current relay.

4. Explain how you would use an ohmmeter to troubleshoot a current relay.

____

____

Sequence to Complete the Lab Task

Starting the Capacitor-Start, Capacitor-Run Hermetic Compressor and Observing the Operation of the Potential Relay

In some HVAC applications where larger compressors are used, three-phase voltage is preferred but may not be available, so single-phase voltage must be used. In this case the amount of starting torque required to start the compressor will be very high, so both a start capacitor and a run capacitor will need to be used to develop this large amount of starting toque. Since a start capacitor and a run capacitor are used, a potential relay will be used to control the start winding and bring it in and out of the circuit. Figure 10–9 shows a diagram of a capacitor-start, capacitor-run compressor. In this diagram you can see the normally closed contacts of the potential relay are wired in series with the start capacitor and the start winding. The coil of the potential relay has very high resistance and is connected across terminals S and C on the compressor. When power is first applied to the compressor, voltage will flow through the normally closed contacts and to the start capacitor and start winding. This will cause current to flow in the start winding that will be out of phase with the current flowing in the run winding, which will cause the maximum amount of starting torque.

When the motor shaft starts to turn and come up to speed, the motor will begin to develop a counter voltage called CEMF between terminals S and C. This voltage will build to a point where it will cause the coil of the potential relay to energize, and cause its normally closed contacts to open. When the normally closed contacts open, they will effectively disconnect the start winding and start capacitor from L1 power.

The run capacitor is connected between the run winding terminal and the start winding, just like in a PSC motor. When the potential relay opens its normally closed contacts, the compressor motor acts like a PSC motor and continues to run. When power is turned off to the compressor, the motor shaft will coast to a stop and the normally closed contacts of the potential relay will return to their closed position and the motor is ready to start again.

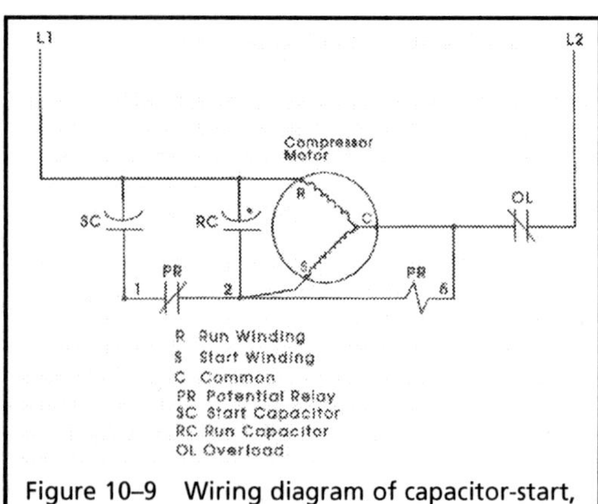

Figure 10–9 Wiring diagram of capacitor-start, capacitor-run compressor that uses a potential relay for control.

1. Your instructor will provide a potential relay that is not connected to a compressor or to a power source. Locate the terminals for the coil of the potential relay at terminals 2 and 5. Use an ohmmeter to check the resistance of coil. Remember, the resistance of the potential relay coil should be very high. Record the resistance of the coil here. __________

2. Use an ohmmeter to measure the resistance of the normally closed contacts of the potential relay, which are located between terminals 1 and 2. Identify if the contacts of the potential relay are normally open or normally closed. __________

3. This section of this lab exercise should be fully supervised by your instructor. Have your instructor initial here when you are ready to complete this part of the exercise. _______

4. In this part of the lab exercise you are going to use a clamp-on ammeter to measure the current in the start winding and run winding of a capacitor-start, capacitor-run hermetic compressor that has its start winding controlled by a potential relay. Be sure to have the power turned off to the motor and place the clamp-on ammeter around one of the wires going to the start winding. When the clamp-on ammeter is in place around the wire for the start winding, turn power on and record the current flowing in the start winding when the motor starts, and again when the motor has run for a few seconds and comes up to speed.
   a. What is the current when voltage is first applied? __________
   b. What is the current when the motor comes up to speed? __________

5. Turn the power to the compressor motor off and let the motor stop. Place the clamp-on ammeter around the winding to the run winding. Start the motor again and allow it to run for 20 seconds. Observe the current in the run winding and record it here.
   a. What is the current level when the motor first starts? __________
   b. What is the current level when the motor has run for 20 seconds? __________

6. Turn the power to the motor off and let the motor stop. Place the clamp-on ammeter around the L1 power line that supplies voltage to both the start winding and the run winding. Start the motor and allow it to run for two minutes. Record the current when the motor first starts. This is called the locked-rotor amperage (LRA). __________

7. Record the motor current after the motor has come up to speed and the start winding has been disconnected by the current relay. This current is called the full-load amperage (FLA). __________

8. Explain what the problem would be if the motor continued to pull LRA values after it first starts and did not return to FLA values.

   ____

   ____

9. Explain how you could use the clamp-on ammeter for troubleshooting a single-phase compressor.

   ____

   ____

10. What would you expect would happen to the compressor motor if the potential relay coil had an open and failed?

____

____

11. What would you expect would happen to the compressor if the contacts of the potential relay were open and would not pass voltage?

____

____

12. Explain how you would use an ohmmeter to test a potential relay to ensure it is operational.

____

____

Checking Out

When you have completed this lab exercise, clean up your area, return all tools and supplies to their proper place, and check out with your instructor. Your instructor will initial here to indicate you are ready to check out. _______

CHAPTER 11

Three-Phase Open Motors and Three-Phase Hermetic Compressors

OBJECTIVES

At the end of this lab exercise you will be able to:

1. Explain the data found on a typical three-phase open motor or hermetic compressor data plate.
2. Measure the resistance of the three windings of a three-phase open motor or hermetic compressor.
3. Understand the function and operation of the internal overload in the hermetic compressor.
4. Measure full-load amperage (FLA) and locked-rotor amperage (LRA) with a clamp-on ammeter.
5. Change the direction of rotation of a three-phase open-type motor.

INTRODUCTION AND OVERVIEW

On commercial installations where three-phase voltage is available, hermetic and open-type motors will usually be three phase. Three-phase motors, also called poly-phase motors, provide high starting torque and excellent running efficiency. These motors are used to drive compressors, pumps, and fans in air-conditioning and refrigeration systems. The air-conditioning and refrigeration technician must be able to wire the three-phase open motors for a change of rotation, a change of voltage, or both, and also be able to troubleshoot and install both hermetic and open-type three-phase motors.

TERMS

Changing rotation
Delta-wired motor
Hermetic compressor
Nine-lead motor
Phase
Six-lead motor
Speed
Three-phase open motor
Three-phase voltage
Torque
Variable-frequency drive (VFD)
Wye-wired motor

MATCHING

Place the letter A–K for the definition from the list that matches with the terms that are numbered 1–11.

Score ________

1. ____ Changing rotation of a motor
2. ____ Delta-wired motor
3. ____ Hermetic compressor
4. ____ Nine-lead motor
5. ____ Phase
6. ____ Six-lead motor
7. ____ Speed
8. ____ Three-phase voltage
9. ____ Torque
10. ____ Variable-frequency drive (VFD)
11. ____ Wye-wired motor

A. One of three-phase electrical power supply lines that provide power on L1, L2, and L3, and a single-phase power supply provides power between L1 and N.
B. A three-phase motor connection where windings are connected in series with each other. The windings are connected in the shape of a triangle or the Greek letter delta.
C. A three-phase motor whose windings are connected in the shape of the letter "Y."
D. The rate of rotation of the shaft of an electrical motor that is measured in revolutions per minute (rpm).
E. A turning force applied to a shaft tending to cause rotation.
F. The process of changing the direction of rotation of a motor's shaft to ensure a fan or pump is rotating in the correct direction.
G. Voltage that is supplied as three separate lines that are identified as L1, L2, and L3. Each phase is 120° out of phase with the other.
H. A compressor used in an HVAC system that is completely sealed. The electric motor and compressor pump are both sealed in a metal container.
I. An electronic system that connects to a single-phase or three-phase AC motor and controls its speed by varying its frequency.
J. A three-phase motor that has six leads brought out of its case for connections.
K. A three-phase AC motor that has nine leads available for connection that allows the motor to be connected as a delta or wye motor, and its windings can be connected for high (480 VAC) or low voltage (240 VAC).

TRUE OR FALSE

Place a *T* or *F* in the blank to indicate if the statement is true or false.

Score ________

1. ____ The direction of rotation of a three-phase motor can be reversed by exchanging any two of its three terminals.
2. ____ One of the windings in a three-phase motor is smaller than the other two so that a phase shift can be created to help the motor get started.
3. ____ The three-phase motor has more starting torque than an equal size single-phase motor.
4. ____ It is possible to change the internal winding connection on a nine-lead three-phase open-type motor so that it can operate on 480 VAC or 240 VAC.
5. ____ It is possible to change the internal windings on a hermetic three-phase compressor so it will operate at a different voltage.

MULTIPLE CHOICE

Circle the letter that represents the correct answer to each question.

Score __________

1. If you change motor terminal T1 to L2 of the supply voltage, and terminal T2 to L1 of the supply voltage, the three-phase motor will:
   a. not run correctly because it must be wired L1 to T1, L2 to T2, and L3 to T3.
   b. change the direction of its rotation.
   c. be able to run on 240 VAC instead of 480 VAC.

2. The three-phase motor does not require any current relays, potential relays, or a centrifugal switch because:
   a. it has three equal windings instead of a start winding.
   b. it is a low-torque motor, and these relays would cause its torque to become too large.
   c. these switches are mainly used to control for overcurrent conditions.

3. The phase shift that is required to create torque for a three-phase motor is:
   a. created by adding capacitors to the motor windings.
   b. found naturally in the three-phase voltage that is used to provide power for the motor.
   c. created by making each of the windings in the three-phase motor slightly different.

4. You need to know how to reconnect a three-phase motor for a change of voltage because:
   a. the voltage rating of the motor that you have in stock is rated 480 VAC, and the motor that it is replacing is rated for 240 VAC.
   b. the motor may need more speed.
   c. some three-phase motors need DC voltage instead of AC voltage.

5. If a three-phase motor runs but draws excessive current, you should suspect:
   a. a faulty centrifugal switch or current relay.
   b. that the motor is running in the wrong direction, and you should exchange any two leads.
   c. that one of the motor windings is open, or one phase of the three-phase voltage is not supplying voltage.

LAB EXERCISE: OBSERVING AND UNDERSTANDING THREE-PHASE MOTORS

Safety for this Lab Exercise

In this lab exercise you have one or more three-phase open and hermetic motors. Some of the motors will be located on a workbench that *will not* be connected to voltage, but you must be aware of the weight of these motors so that they do not roll off the bench. You will be requested to measure the resistance in the windings of these motors and take data from their data plates.

Your instructor will provide additional three-phase open and hermetic compressors that are connected to a power source so that you can apply voltage to them and observe their operation. Any time you will have voltage applied to the motors or compressors, you will need to be supervised by your instructor. You need to be aware of the voltage supplied to the terminals or components of these compressors so that you do not receive an electrical shock.

Tools and Materials Needed to Complete the Lab Exercise

Your instructor will provide a three-phase open-type motor and hermetic compressor that is removed from an HVAC or refrigeration system and placed on a bench so that you can measure the resistance of its windings, and take data from its data plate. You will need additional three-phase open motors and hermetic compressors. These motors need to be in an operational system so that you can connect power to them and run them. When the motors and compressors are running, you will be requested to take current measurements. You will need a clamp-on ammeter for measuring the starting current, which is also called locked-rotor amperage (LRA), and the full-load amperage (FLA). Your instructor will provide a three-phase hermetic compressor that has its internal overloads open, and one with an external overload. These units will be used to measure resistance of the winding to determine that the internal overload contacts or the external overload contacts are open. These compressors should not be connected to power so students can make a resistance check of the windings.

References to the Text

Refer to Chapter 11 in the textbook for additional information. You may need to read sections of the chapter again to help you understand the material in this exercise.

Sequence to Complete the Lab Task

Taking Data from the Three-Phase Open-Type Motor and Hermetic Compressor's Data Plate

The basic theory of operation is the same for both the hermetic and the open-type three-phase induction motor. The major difference is that the hermetic motors generally have only three terminals. These terminals are called T1, T2, and T3. Open-type motors will usually bring the end of each winding out to a terminal board or inspection plate. Figure 11–1 shows the terminals of a three-phase hermetic compressor, and Figure 11–2 shows the windings of a three-phase open-type motor.

Most three-phase open motors have 9 or 12 terminals. These are identified as T1, T2, T3, etc. The three-phase motor has three separate windings that may consist of one or more sets of coils connected together that are powered by three separate phases of voltage. Since each phase of the three-phase voltage is separated by 120° and the three windings are displaced in the three-phase motor, no starting device or start winding is required to get the magnetic field to rotate in the stator to provide the strong starting torque. Since the rotor is a "squirrel cage" rotor, a magnetic

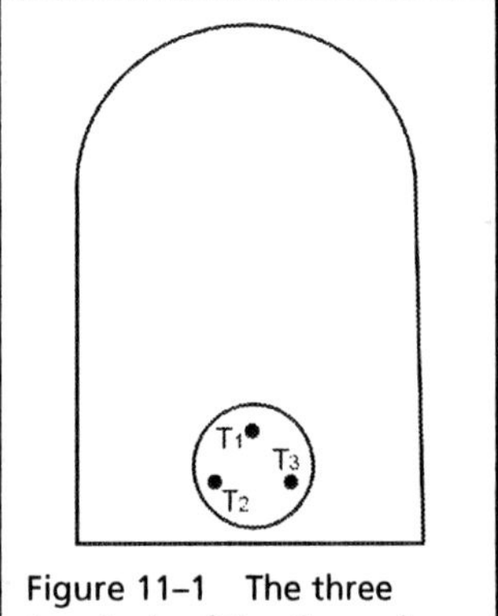

Figure 11–1 The three terminals of the three-phase hermetic compressor.

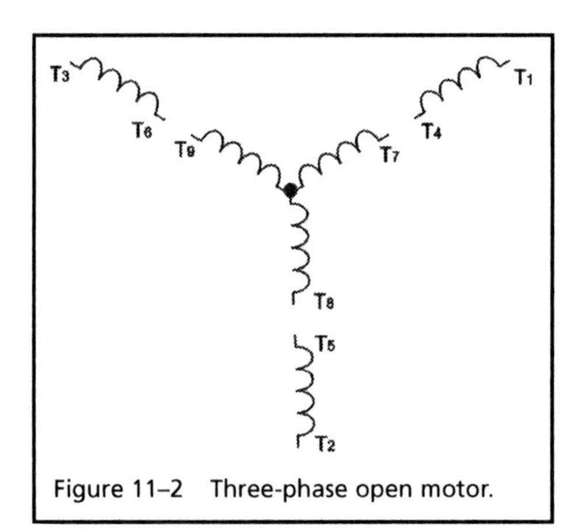

Figure 11–2 Three-phase open motor.

field is induced into it just as in the single-phase induction motors. This means that as soon as three-phase power is provided to the windings, the rotor begins to move.

The three windings of the three-phase motor are equal in magnetic strength. This gives the motor good efficiency and makes it more durable. In an open-type motor, these windings can be wired in a star (wye) configuration or delta (Δ) configuration. The star- and delta-wired motors are similar to one another in operational performance. The major difference is that the wye-wired motor draws less starting current and provides a little less starting torque. If one of the three windings has an open, it will not draw current and the motor will not be able to start. When power is applied to the motor that has an open in one of its windings, the motor will hum, but will not start, and if it is allowed to try to start, it will eventually blow the fuses protecting the other two windings. If a motor has two windings open, the motor will not draw any current and it will not make any noise or try to start.

1. Your instructor will provide a three-phase open-type motor and hermetic compressor on a workbench so that you can take data from their data plates. Be sure that this motor and compressor is not connected to a power source and that it cannot be turned on. Find the data plate on the compressor and record the following data if it is present on the data plate. Have your instructor check your information for accuracy.
   a. HP __________
   b. Voltage __________
   c. Cycles/hertz __________
   d. FLA __________
   e. LRA, if provided __________
2. Find the data plate on the three-phase open-type motor and record the following data if it is present on the data plate. Have your instructor check your information for accuracy.
   a. HP __________
   b. Voltage __________
   c. Cycles/hertz __________
   d. FLA __________
   e. LRA, if provided __________
   f. Direction of shaft rotation, if specified __________
   g. Explain why it is important to locate the data plate of a compressor- or open-type motor and record the information if you are changing it out.

      ____

      ____

Sequence to Complete the Lab Task

Measuring the Resistance of the Three-Phase Compressor

1. Use the diagram shown in Figure 11–3 to measure the resistance the windings of the three-phase hermetic compressor. You should notice from this diagram that the three windings are equal in size and resistance, and are identified as T1, T2, and T3. You can measure the

resistance of windings 1 and 2 by putting one terminal of the ohmmeter on terminal T1 and the other on terminal T2. To calculate the resistance of each winding, you would divide the total resistance you have measured between T1 and T2 by two. Record the resistance between T1 and T2. __________

2. Measure the resistance between T1 and T3 and record it here. __________

3. Measure the resistance between T2 and T3 and record it here. __________

4. Was the resistance of all three measurements about equal? __________

5. What would you suspect if the resistance between T1 and T2 was 30 ohms, and the resistance between T1 and T3 was infinite, and the resistance between T2 and T3 was infinite?

____

____

6. Explain how you could use the resistance measurements of a three-phase motor to find one or more open windings.

____

____

Sequence to Complete the Lab Task

Measuring the Resistance of the Three-Phase Open-Type Motor

Use the diagram shown in Figure 11–2 to measure the resistance of the windings of the three-phase open-type motor. You should notice from the diagrams that the three windings are actually made of six separate windings. Three of the windings are found between terminals 1–4, 2–5, and 3–6, and the final three windings are connected at a common point that you cannot get to, so you will measure the resistance of these three windings from terminals 7–8, 8–9, or 9–7. These windings can be wired in series as shown in Figure 11–2, or they can be connected in parallel. If any of these windings are open, the motor will not operate correctly. In this section of the exercise you will be provided a three-phase open-type motor that is not connected to power, so you can safely check the resistance of the windings.

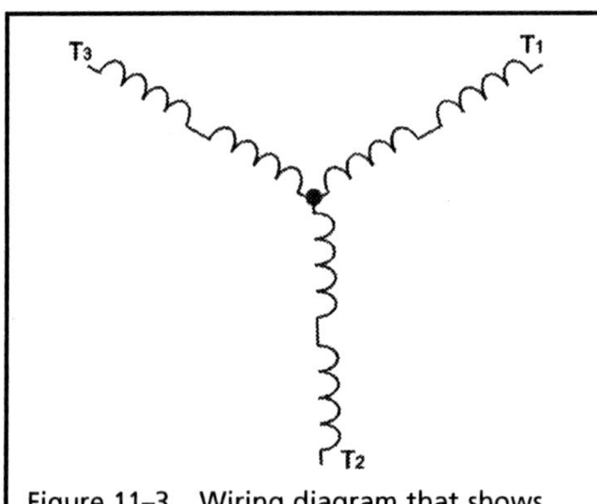

Figure 11–3 Wiring diagram that shows the windings of a three-phase hermetic compressor.

1. Locate the terminals T1 and T4. Measure and record the resistance of the winding.

   ____

2. Measure the resistance between T2 and T5, and record it here. __________

3. Measure the resistance between T3 and T6, and record it here. __________

4. Was the resistance of all three measurements about equal? __________

5. What would you suspect if the resistance between any two of the above terminals were infinite?

   ____

   ____

6. The next step is to measure the resistance of the three windings that have the common point.

   a. Measure and record the resistance between T7 and T8. __________

   b. Measure and record the resistance between T8 and T9. __________

   c. Measure and record the resistance between T9 and T7. __________

   d. Explain what the measurements would be between T7, T8, and T9 if one of these windings had an open in it.

      ____

      ____

Sequence to Complete the Lab Task

Checking the Internal and External Overload for the Hermetic Compressor

In this part of the lab exercise you will learn about the internal overload for the hermetic compressor. The internal overload is buried in the motor winding to detect excessive heat buildup that can damage the windings. The internal overload consists of a set of normally closed contacts that are wired in series with the common point in the three-phase motor where all three windings come together. If the motor windings experience excessive heat buildup, the normally closed contacts of the internal overload open and stop the current flow through any of the three windings. When you are testing the windings, you will be able to tell if the contacts of the internal overload are open or closed. If you come upon a three-phase compressor that will not start and does not draw any current, you should disconnect all power to the compressor and test the windings for resistance. If the resistance measurement in all three windings indicate the windings are open, you should suspect the internal overload and allow the compressor to cool down for several hours. Sometimes you must wait overnight for the overloads to reset. Sometimes the windings become too hot because the HVAC or refrigeration system is low on refrigerant, and the hermetic compressor relies on some amount of excess refrigerant to come back to the motor windings to help cool them. If the overload contacts open and will not reset, the compressor will need to be removed and replaced.

1. Your instructor will provide you with a hermetic compressor that has its overloads open. This compressor *should not be connected to a power source*. Make a resistance measurement of the windings and record the results below.

   a. Measure resistance from T1 to T2. __________

   b. Measure resistance from T1 to T3. __________

c. Measure resistance from T2 to T3. __________

d. Explain why you think the internal overload is open in this compressor.

____

____

2. Some compressors have an external overload instead of an internal overload. The external overload is mounted so that it makes contact with the compressor where the windings are located and where the over temperature conditions will occur. It is important that the external overload makes physical contact with the metal of the compressor so that heat can be transferred easily. The overload will heat up two ways—first, from the current flowing through it to the motor windings, and second, from any heat that comes through the compressor metal container. Checking the external overload is much simpler than checking the internal overload, because you can get to the terminals of the overload and check them for resistance. In this exercise your instructor will provide a compressor with an external overload. Locate the overload and make sure power is turned off to the compressor. Use an ohmmeter to measure the resistance of the contacts of the overload and indicate if the contacts are open or closed. __________

3. If the external overload trips, you must wait for the compressor motor to cool down before you try to start it again. If you find an external overload that has tripped when you are troubleshooting, explain what you would check.

____

____

Sequence to Complete the Lab Task

Starting the Three-Phase Compressor and Measuring the Starting Current and the Full-Load Current (FLA) with a Clamp-on Ammeter

1. This section of this lab exercise should be fully supervised by your instructor. Have your instructor initial here when you are ready to complete this part of the exercise. _______

   Your instructor will provide a three-phase compressor that is connected to a power source. During this exercise you will start the compressor several times and measure the current each winding draws. The current motor draws when it is running is called the full-load amperage (FLA) for the compressor. This amount of current will vary slightly as the load on the compressor increases or decrease.

2. Place the clamp-on ammeter around the wire connected to the T1 terminal. Apply power to the compressor for several minutes and measure the current to this winding and record it here. __________

3. Place the clamp-on ammeter around the wire connected to the T2 terminal. Apply power to the compressor for several minutes and measure the current to this winding and record it here. __________

4. Place the clamp-on ammeter around the wire connected to the T3 terminal. Apply power to the compressor for several minutes and measure the current to this winding and record it here. __________

5. Comment on the amount of current that each winding was drawing.

 ____

 ____

6. Is the current balanced across each winding? __________

7. In the next step you are going to measure the locked-rotor amperage (LRA) for the compressor. You will need to set your clamp-on ammeter so that it measures the maximum current draw when the motor starts and locks on this value. Turn power off and place the clamp-on ammeter around the wire connected to any of the three windings. Turn power on to the motor and observe the current level when the motor first starts. Record the current when the motor first starts. __________

8. You should notice the LRA is much higher than the FLA, but the LRA only lasts for a few seconds. For this reason you must use time delay fuses that are rated for the FLA. The time delay that is built into the fuse is set for approximately 20 to 30 seconds, which allows plenty of time for the motor to start and draw the higher LRA and then return to the FLA rating. What would happen to the fuses if the motor continued to draw LRA for more than 30 seconds?

 ____

 ____

Sequence to Complete the Lab Task

Reversing the Rotation of the Three-Phase Motor

At times you will install a three-phase pump or fan and the shaft of the motor will be turning in the wrong direction. As an installer, you will be expected to check the rotation and make changes as needed. The three-phase motor can have its rotation reversed by reversing any two motor leads to the incoming voltage lines. For example, the T1 motor line that is connected to incoming voltage line L1 can be changed and connected to L2, and T2 motor line that is connected to incoming voltage line L2 can be changed and connect to L1. (The same result can be obtained by changing T2 and T3, or T1, and T3.)

Direction of rotation of an open-type motor is described as clockwise (turning to the right), or as counter-clockwise (turning to the left), and it is determined by observing the shaft rotation from the end opposite the shaft. In other words, you would stand behind the motor and watch the direction the shaft turns.

Another time you may need to reverse the rotation of a three-phase motor is when a three-phase compressor is stuck and will not start. This occurs sometimes when an air-conditioning system has been turned off during the winter and not run for several months. When you check out this problem you will find that all three phases of voltage are provided to the motor terminals, all three windings have continuity and are drawing current, but the compressor does not want to start, and it just hums and draws locked-rotor amperage (LRA). If you allow the motor to try to start for more than one or two seconds, the fuses protecting the motor will blow. One way to try to get the compressor to start is to reverse the rotation of the three-phase motor. Since this is a hermetic compressor, the direction of rotation does not matter, so you can change it and not cause harm to the compressor. When the compressor is started in the opposite direction, sometimes it will free up and start.

1. This section of this lab exercise should be fully supervised by your instructor. Have your instructor initial here when you are ready to complete this part of the exercise. _______

   Be sure you are aware of electrical safety and have all the electrical terminals covered and the door of the disconnect closed when you apply three-phase voltage to the motor. You must also be aware of the motor shaft and make sure it is clear to rotate without causing a danger to anyone or any equipment. Also, make sure the motor is mounted securely so it does not move about.

   Your instructor will provide a three-phase open-type motor that is connected to three-phase disconnect and securely mounted in a piece of HVAC equipment. You will be requested to check the rotation of its shaft and make changes to the motor leads to the incoming voltage lines to cause it to reverse its rotation.

2. Turn power on to the motor by closing the disconnect switch. Allow the motor to run for several seconds and then turn it off and allow it to coast to a stop. Stand behind the motor at the end opposite where the shaft comes out and observe if the motor is turning clockwise or counter-clockwise. Record your observations here.

   The motor shaft is turning in the __________ direction.

3. Have your instructor ensure that power is off at your disconnect and that you can open the door of the disconnect to measure the voltage at the terminals at the bottom of the disconnect. Have your instructor initial here to indicate that voltage is turned off and it is safe to work in the disconnect. _______

4. After you have determined the voltage is turned off, you can take the motor lead for terminal 1 and remove it from terminal T1 of the fuse disconnect. Also take the motor lead for terminal 2 and remove it from terminal T2 of the fuse disconnect. Connect the motor lead for terminal 1 and connect it under the T2 terminal of the fuse disconnect, and take motor lead for terminal 2 and connect it under the T1 terminal of the fuse disconnect. Have your instructor check your wiring and initial this space to indicate it is okay. _______

5. After you have determined the motor is connected for the reverse rotation, close the disconnect door and turn power on to the disconnect again for several seconds and check the direction of the motor's shaft rotation. Record your observations. If your motor does not change the direction of its shaft rotation, call your instructor immediately.

   The motor shaft is now turning in the __________ direction.

Checking Out

When you have completed this lab exercise, clean up your area, return all tools and supplies to their proper place, and check out with your instructor. Your instructor will initial here to indicate you are ready to check out. _______

CHAPTER 12

Relays, Contactors, Solenoids, Motor Starters, and Overcurrent Controls

OBJECTIVES

At the end of this lab exercise you will be able to:

1. Measure the resistance of a relay, contactor, and motor starter coil.
2. Identify the normally open contacts, normally closed contacts, and coil of the relay, and explain their operation and functions.
3. Explain the data found on a typical relay, contactor, and motor stater data plate.
4. Explain the operation of a motor starter and its overloads.
5. Explain how to reset an overload on a motor starter.

INTRODUCTION AND OVERVIEW

Relays, contactors, and motor starters are used in HVAC systems to control loads such as electric heating elements, fan motors, and compressors, and turn them on and off. The relay, contactor, and motor starter each have a coil and one or more sets of contacts. The coil is made of multiple turns of wire that becomes a very strong magnet when voltage is applied to it and current flows through it. The magnetic field of the coil causes the contacts to move and change state, which means normally open contacts will go closed, and normally closed contacts will go to their open state. The moving part that moves the contacts is called the armature. You can check in the text to see more details about the types of armatures used in relays, contactors, and motor starters. Normally open contacts have high resistance and are open when no power is applied to the coil, and normally closed contacts have low resistance and are closed when no power is applied to the coil. When voltage is turned off to the coil, current stops flowing through it, its magnetic field stops, and the springs on the contacts cause them to return to their original state. The amount of voltage and current that is used to operate the coil is rather small compared to the voltage and current that flows through the contacts to the motors or other loads. Typically, the coils for relays and motor starters are in the control circuit of a system, and the contacts are in the load circuit with the motors and other loads. Contactors are similar to relays in that they have a coil and one or more sets of contacts, but the contacts or the contactor are rated for loads that are larger than relays. By definition, a contactor has contacts that are rated for more than 15amps, and a relay has contacts that are rated for less than 15 amps.

TERMS

Armature
Bell-crank type relay
Clapper type relay
Coil
Control circuit
Double-pole, double-throw (DPDT)
Double-pole, single-throw (DPST)
Hold-in current
Horizontal action
Inrush current
Load circuit
Normally closed contacts
Normally open contacts
Pull-in current
Relay
Relay coil
Relay contacts
Single-pole, double-throw (SPDT)
Single-pole, single-throw (SPST)
Transformer
Vertical-action relay

MATCHING

Place the letter A–S for the definition from the list that matches with the terms that are numbered 1–19.

Score __________

1. ____ Armature
2. ____ Bell-crank type relay
3. ____ Clapper type relay
4. ____ Coil
5. ____ Control circuit
6. ____ Double-pole, double-throw (DPDT)
7. ____ Double-pole, single-throw (DPST)
8. ____ Hold-in current
9. ____ Inrush current
10. ____ Load circuit
11. ____ Normally closed contacts
12. ____ Normally open contacts
13. ____ Pull-in current
14. ____ Relay
15. ____ Relay coil
16. ____ Relay contacts
17. ____ Single-pole, double-throw (SPDT)
18. ____ Single-pole, single-throw (SPST)
19. ____ Vertical-action relay

A. The part of an electrical diagram that contains the switches that control when the loads (motors or electrical resistance coils) turn on or off. In an HVAC system, this circuit usually contains the temperature, pressure, or flow switches.
B. A relay that has a coil mounted above its armature and a set of movable contacts and stationary contacts that are mounted in the horizontal position. When the coil becomes energized, it pulls the armature upward.
C. A switch consisting of two sets of normally open contacts. Both sets of contacts are actuated together.
D. The moving part of a magnetic circuit, such as the rotating part of a motor, a generator, or the movable iron part of a relay.
E. A relay that has its contacts mounted above its coil. When the relay coil is energized, the armature is pulled upward and causes the contacts to change from open to closed.
F. The amount of current used to pull in a relay coil.

G. This relay uses a yoke that is pulled to the magnetic coil when the coil is energized. The yoke moves a pushrod, and this physical movement of the armature of the clapper forces the pushrod and movable contacts upward.

H. A set of contacts on a relay or switch that are closed and have low resistance when the switch or relay is in the de-energized state.

I. A part of a relay or solenoid that is made by tightly winding a long piece of wire into loops. When current flows through the wire, it creates a strong magnetic field in the coil that can be used to open and close contacts in a relay.

J. The part of a relay that is made of a coil of wire through which current flows and creates a magnetic field. The magnetic field causes the contacts to change position.

K. A set of contacts that have one input and one output.

L. The part of an air-conditioning or heating circuit that has large loads such as the compressor in it. The control circuit has the thermostat and other controls in it.

M. A switch consisting of two sets of contacts, and each set of contacts has a normally open and normally closed contact that is connected to a common "C" point. Both sets of contacts are actuated together.

N. A set of contacts on a relay or switch that are open and have high resistance when the switch or relay is in the de-energized state.

O. The part of the relay that has the electrical contacts (either normally open or normally closed).

P. The amount of current flowing through a relay coil immediately after pull-in current. This current causes a set of relay contacts to remain energized.

Q. An electrical device that consists of a single coil and one or more sets of normally open or normally closed contacts.

R. Large current caused when voltage is first applied to a motor or other inductive load. This current will return to normal full-load current several seconds after the motor starts and comes up to full rpm.

S. A single set of contacts that have a common point and can be switched to either of two positions.

TRUE OR FALSE

Place a *T* or *F* in the blank to indicate if the statement is true or false.

Score __________

1. ____ Plug-in relays are used in air-conditioning and refrigeration systems because they are easy to change when they become faulty.

2. ____ The main difference between a relay and a contactor is that the contactor usually has more contacts.

3. ____ A solenoid is similar to a relay in that it has a coil that becomes magnetic and moves the plunger of a valve.

4. ____ It is possible to change normally open contacts to normally closed contacts in the field for some relays.

5. ____ For a relay to operate correctly, its contacts must close first to provide current to its coil.

6. ____ The heater is the part of the motor starter overload that opens and interrupts current flow.

7. ____ The overload contacts on a motor starter are normally open and interrupt current when an overload condition is detected.

8. ____ The main job of the circuit breaker is to protect the circuit against short circuits and the entire circuit against overcurrent, rather than to protect individual motors against overcurrent.

9. ____ A fused disconnect provides a location for mounting fuses for the system and also provides a means of disconnecting main power from the circuit.

10. ____ An overload device may be mounted internally in a compressor or externally to detect heat on the compressor shell.

MULTIPLE CHOICE

Circle the letter that represents the correct answer to each question.

Score __________

1. The normally closed contacts of a relay pass current:
   a. when the coil is not energized.
   b. when the coil is energized.
   c. at all times, whether or not the coil is energized.

2. A current relay is a:
   a. special relay designed to control only current.
   b. relay that is designed to be used as a relay or a contactor.
   c. relay that is designed to start single-phase compressors.

3. A solenoid is a:
   a. valve that is controlled (opened and closed) by gravity and springs.
   b. reversing relay that is used to start heat pumps.
   c. valve that is controlled (opened and closed) by springs and a magnetic field produced by a coil.

4. A reversing valve is a special:
   a. type of solenoid used in heat pumps.
   b. relay used to reverse the direction in which the compressor motor runs.
   c. valve that can be used as either a fill valve or a drain valve for a heat pump.

5. An SPDT relay has:
   a. one set of contacts that connects to two output terminal points.
   b. two sets of contacts that connect to one terminal point.
   c. one set of contacts and one output terminal point.

6. A motor starter:
   a. is just a larger version of a relay, since it has only a coil and contacts.
   b. has a coil and contacts like a relay or contactor, and it also has overloads to provide overcurrent protection.
   c. is more like a circuit breaker than a relay in that it can detect excessive heat buildup on the outside of the compressor motor through its heaters.

7. A dual-element fuse:
   a. can detect slow overcurrent and large short-circuit currents.
   b. can be used two times before it must be replaced.
   c. has two elements, so it can be used in single-phase circuits to protect both L1 and L2 supply voltage lines.

8. The overload device on a motor starter has a heater:
   a. to detect excess current and overload contacts to interrupt the control circuit current.
   b. to detect excessive current and overload contacts that directly interrupt the large load current to the motor.
   c. that interrupts control circuit current and overload contacts that directly interrupt the large load current to the motor.

9. The motor starter has auxiliary contacts that:
   a. are connected in parallel with a start switch to seal in the circuit when it is operating correctly, and ensure the coil remains de-energized if the overload contacts trip.
   b. are connected in series with a start switch to seal in the circuit when it is operating correctly, and ensure the coil remains de-energized if the overload contacts trip.
   c. can switch current to an auxiliary motor such as a condenser fan at the same time the main contacts supply current to the compressor motor.

10. Heaters for motor starters are available in different sizes so that they can:
    a. provide the correct amount of current protection to match the motor that is connected to the motor starter.
    b. match the voltage rating of the fuses used in the circuit.
    c. match the temperature rating of the ambient air in which the motor will operate.

LAB EXERCISE: OBSERVE AND UNDERSTAND RELAYS, CONTACTORS, AND MOTOR STARTERS

Safety for this Lab Exercise

In this lab exercise you have several relays, contactors, and motor starters that are located on a workbench that *will not* be connected to voltage. You will be requested to take data from their data plates and measure the resistance in their coils and contacts so that you can identify them. You will also be provided a relay, contactor, and motor starter that are connected to loads in an operating HVAC system, and you will observe their operation and test them for voltage and current.

If the relay, contactor, or motor starter is connected to an operating circuit, you must be aware that they will have exposed electrical terminals where you can receive a severe electrical shock if you come into contact with them. Be sure to work safely around these devices when you are taking voltage and current measurements. It is also important that you turn off power to any circuit where you are making resistance tests for relays, contactors, or motor starters. Also, if the motor is turning a fan or other mechanical device, you must be aware to stay clear of all moving parts or mechanical parts.

Tools and Materials Needed to Complete the Lab Exercise

Your instructor will provide a number of relays, contactors, and motor starters from an HVAC or refrigeration system and place them on a bench so that you can observe their operation and measure the resistance of their coils and contacts. You will also take data from their data plate.

Your instructor will provide additional relays, contactors, and motor starters that are connected to motors and other loads so that you can turn them on and off and observe their operation and take voltage and current measurements.

References to the Text

Refer to Chapter 12 in the textbook for additional information. You may need to read sections of the chapter again to help you understand the material in this exercise.

Additional Information for Relays, Contactors, and Motor Starters

Relays and contactors may have one or more sets of normally open and normally closed contacts. Figure 12–1 shows pictures of typical relays and contactors used in HVAC systems, and Figure 12–2 shows diagrams of the types of contacts available for relays and contactors. You can see in the diagrams that some relays and contactors have normally open contacts, and some have a combination of normally open and normally closed contacts. The contacts on a relay, contactor, or motor starter have both a voltage rating and a current rating. The voltage rating for contacts will be 300 volts or 600 volts. If the contacts are used in a 120- or 240-volt circuit, the contacts rated for 300 volts are used. If the contacts are used in a circuit that has 480 volts, the contacts must be rated for 600 volts. The current rating will be listed in amperes or in horsepower (HP), and should always be rated higher than the load current they are controlling. Single-pole, single-throw contacts are shown in Figure 12–2a; double-pole, double-throw contacts are shown in Figure 12–2b; single-pole, double-throw contacts are shown in Figure 12–2c; and double-pole, double-throw contacts are shown in Figure 12–2d. You can test a relay, contactor, or motor starter contacts with an ohmmeter to determine if they are normally open or normally closed.

Relays and contactors have coils that are rated to operate on AC and DC voltages. Typically, relays used in HVAC and refrigeration systems will be controlled by AC voltage. Typical coil voltages for relays and contactors are 24 VAC, 120 VAC, and 240 VAC. Most HVAC and refrigeration systems use 24 VAC and 120 VAC. You must match the coil to the voltage rating of the circuit it is operating in. If you replace a relay, you must ensure that its contact's voltage and current rating is larger than its load, and its coil's voltage rating must match the voltage of the circuit it is used in.

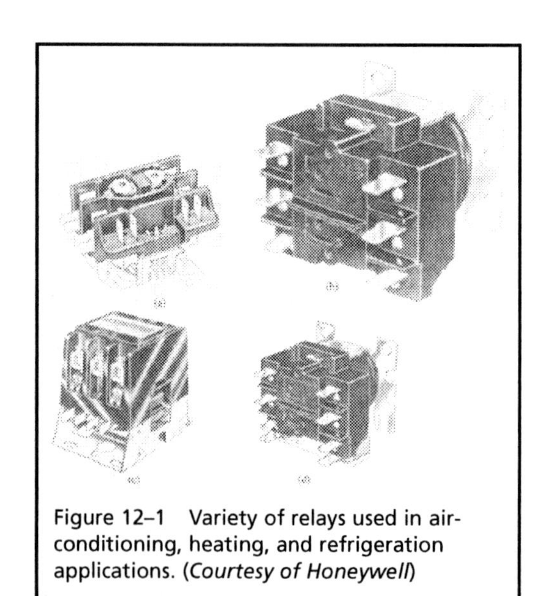

Figure 12–1 Variety of relays used in air-conditioning, heating, and refrigeration applications. (*Courtesy of Honeywell*)

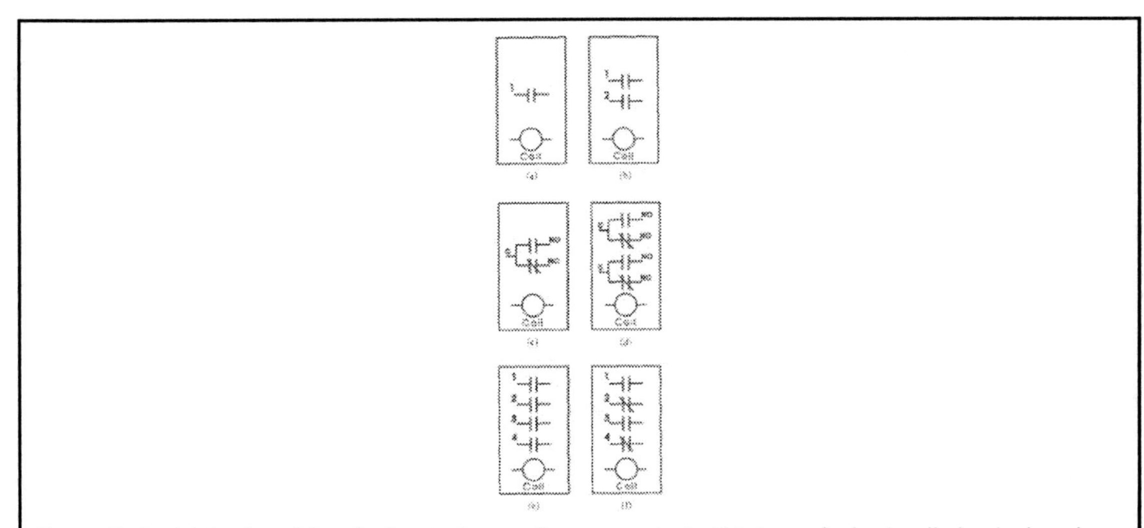

Figure 12–2 (a) A relay with a single set of normally open contacts. This type of relay is called a single-pole, single-throw (SPST) relay. (b) A relay with two individual sets of contacts. This relay is called a double-pole, single-throw (DPST) relay. (c) A relay with two sets of contacts that are connected at one side at a point called the common (C). The output terminals are identified as normally open (NO) and normally closed (NC). This type of relay is called single-pole, double-throw (SPDT). (d) A relay with two sets of SPDT contacts. This relay is called a double-pole, double-throw (DPDT) relay. (e) A relay with multiple sets of normally open contacts. (f) A relay with a combination of normally open and normally closed contacts.

On larger HVAC and refrigeration systems that operate on three-phase voltage, a motor starter is used to protect the compressor motor or fan and pump motors against overcurrent. The motor starter has a coil and multiple contacts, and an overload device that can sense excess current flowing through the device into a motor. The motor starter is shown in Figure 12–3, and the electrical diagram for the motor starter is shown in Figure 12–4. The overload device is made of two parts: the heater, which is a device that creates heat when current flows through it, and the normally closed overload contacts that open when the current level becomes too high. In the diagram in Figure 12–4, you can see the overload heaters are connected in series with the motor windings, and they sense the amount of current flowing to the motor windings and the normally closed overload contacts are wired in series with the motor starter coil. If the motor starter senses too much current flow to the motor windings, the overload contacts open and de-energize the coil of the motor starter. If the overload trips, you will have to manually reset it when you discover it is tripped. You should always check the motor current with a clamp-on ammeter to ensure the motor is pulling the proper amount of current. If the motor is pulling overload current, you will have to fix this problem before you reset the motor starter.

Some pumps and fans are started and stopped with a start and stop button. The motor starter has an extra set of contacts called auxiliary contacts, and these contacts are wired in parallel with the start push button. When the normally open push button is depressed, voltage flows to the motor starter coil. When the coil energizes, it closes the auxiliary contacts, which "seals in" the circuit around the start button, which will return to its normally open position when the person starting the system takes their finger off the button. The auxiliary contacts will open when the stop push button is depressed or the normally closed overload contacts trip, and be ready for the next time the motor is ready to be started.

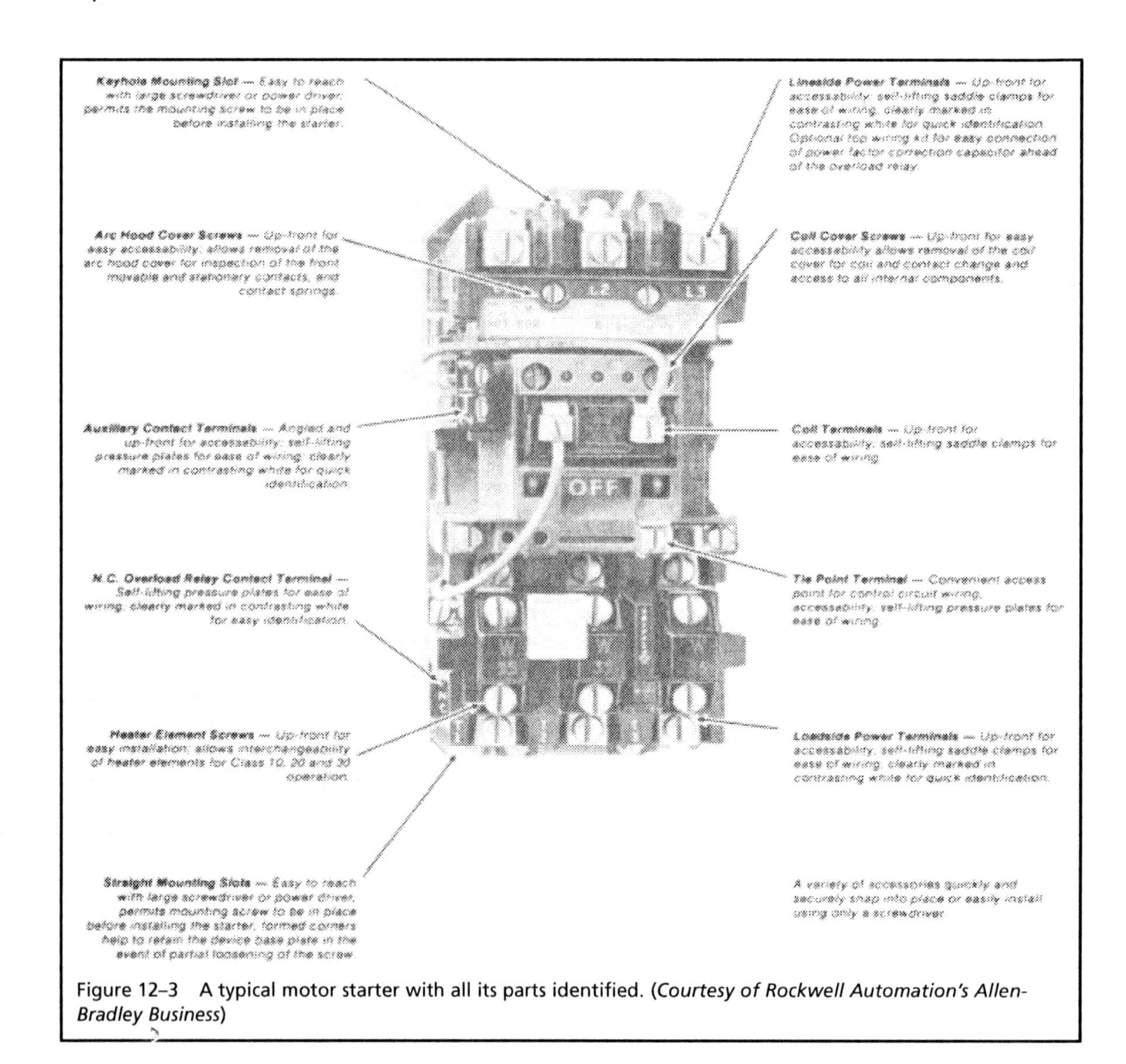

Figure 12–3 A typical motor starter with all its parts identified. (*Courtesy of Rockwell Automation's Allen-Bradley Business*)

Sequence to Complete the Lab Task

Taking Data from Relays, Contactors, and Motor Starter Data Plates

Your instructor will provide a number of relays, contactors, and motor starters on a workbench so that you can take data from the data plate. Be sure that these components are not connected to a power source and that it cannot be turned on. Find the data plate on each and record the following data if it is present on the data plate. Have your instructor check your information for accuracy.

1. Check the relay for the following data:
   a. Coil voltage __________
   b. Contact voltage rating __________
   c. Contact amperage or HP rating __________
   d. Number of NO contacts __________
   e. Number of NC contacts __________

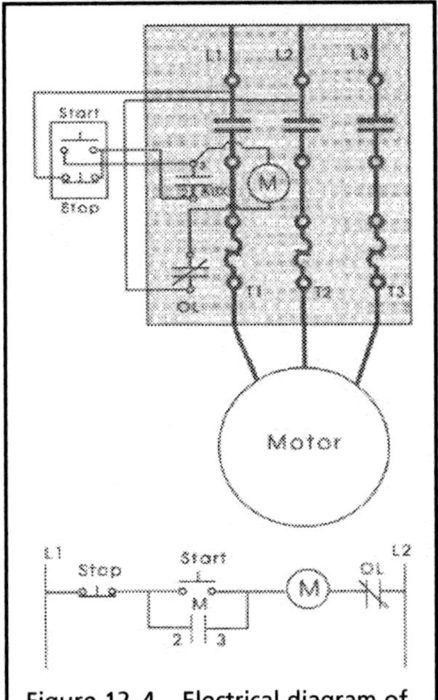

Figure 12–4 Electrical diagram of a motor starter and control circuit.

2. Check the contactor for the following data:
   a. Coil voltage __________
   b. Contact voltage rating __________
   c. Contact amperage or HP rating __________
   d. Number of NO contacts __________
   e. Number of NC contacts __________

3. Check the motor starter for the following data:
   a. Coil voltage __________
   b. Contact voltage rating __________
   c. Contact amperage or HP rating __________
   d. Current rating of the overloads __________
   e. Number of NO contacts __________
   f. Number of NC contacts __________

4. Explain why it is important to locate the data plate of a relay, contactor, or motor starter, and record the information if you are changing it out.

   ____

   ____

Sequence to Complete the Lab Task

Measuring the Resistance of the Coil and Contacts of a Relay and Contactors

Your instructor will provide a number of relays and contactors on a workbench so that you can measure the resistance of their coils and contacts. You will be asked to determine if the contacts are normally open or normally closed. Be sure that these components are not connected to a power source and that they cannot be turned on. Have your instructor check your information for accuracy.

1. Check relay #1 for the following:
   a. Coil resistance __________
   b. Contact set #1 resistance __________ Are these NO or NC? __________
   c. Contact set #2 resistance __________ Are these NO or NC? __________
   d. Contact set #3 resistance __________ Are these NO or NC? __________
2. Check relay #2 for the following:
   a. Coil resistance __________
   b. Contact set #1 resistance __________ Are these NO or NC? __________
   c. Contact set #2 resistance __________ Are these NO or NC? __________
   d. Contact set #3 resistance __________ Are these NO or NC? __________
3. Check relay #3 for the following:
   a. Coil resistance __________
   b. Contact set #1 resistance __________ Are these NO or NC? __________
   c. Contact set #2 resistance __________ Are these NO or NC? __________
   d. Contact set #3 resistance __________ Are these NO or NC? __________
4. Check contactor #1 for the following:
   a. Coil resistance __________
   b. Contact set #1 resistance __________ Are these NO or NC? __________
   c. Contact set #2 resistance __________ Are these NO or NC? __________
   d. Contact set #3 resistance __________ Are these NO or NC? __________
5. Check contactor #2 for the following:
   a. Coil resistance __________
   b. Contact set #1 resistance __________ Are these NO or NC? __________
   c. Contact set #2 resistance __________ Are these NO or NC? __________
   d. Contact set #3 resistance __________ Are these NO or NC? __________

Sequence to Complete the Lab Task

Measuring the Resistance of the Coil and Contacts of a Motor Starter and Identifying the Overload and Heater

Your instructor will provide a motor starter on a workbench so that you can measure the resistance of its coil and contacts and look it over and observe the parts of the motor starter. You will be asked to determine if the contacts on the motor starter are normally open or normally closed. You will also be requested to remove the heater element and identify its size. Your instructor will show you where the auxiliary contacts and the overload contacts are located, and how the heater element can trip the overload contacts. Be sure that these components are not connected to a power source and that it cannot be turned on. Have your instructor check your information for accuracy.

1. Check motor starter #1 for the following:
    a. Coil resistance __________
    b. Contact set #1 resistance __________ Are these NO or NC? __________
    c. Contact set #2 resistance __________ Are these NO or NC? __________
    d. Contact set #3 resistance __________ Are these NO or NC? __________
    e. Auxiliary contacts resistance __________ Are these NO or NC? __________
    f. Overload contacts resistance __________ Are these NO or NC? __________
    g. Identify the heater coil size. __________
    h. What is the current rating of the heating element? __________

Sequence to Complete the Lab Task

Observing the Operation of a Relay or Contactor

1. This section of this lab exercise should be fully supervised by your instructor. Have your instructor initial here when you are ready to complete this part of the exercise. _______
   In this part of the lab exercise your instructor will provide an HVAC system that has an evaporator fan (furnace fan) that is controlled by the fan relay. You will turn the thermostat fan switch to the on position to energize the fan relay coil, which will cause its contacts to close. Your instructor will place the leads of a voltmeter across the terminals where the relay coil is connected and you can place a clamp-on ammeter around one of the leads that connects the fan motor to the fan relay. You will observe the voltmeter to see when voltage is present at the coil and observe when the contacts close and cause the fan motor to run. When the coil becomes energized and closes the relay contacts you will observe the current measurement that is shown on the clamp-on ammeter.

2. Turn the fan switch on the thermostat to the on position. What is the voltage at the coil?
   ____

3. When the coil receives voltage, what happens to the contacts?
   ____
   ____

4. When the contacts go closed what is the amount of current flowing to the fan? __________

5. Use this space to discuss the operation of the fan relay. Explain the operation and include what happens first, second, third, etc.

____

____

6. How will the information in the previous question help you troubleshoot a relay?

____

____

7. Explain what loads you would expect to find connected to a contactor instead of a relay.

____

____

Sequence to Complete the Lab Task

Observing the Operation of a Motor Starter

1. This section of this lab exercise should be fully supervised by your instructor. Have your instructor initial here when you are ready to complete this part of the exercise. _______

2. Your instructor will provide a three-phase motor that is controlled by a motor starter. The motor starter will be turned on by a normally open start push-button switch, and turned off by the normally closed stop push-button switch. Your instructor will place the leads of a voltmeter across the terminals of the motor starter coil. You will place a clamp-on ammeter around one of the wires that connects the motor to the motor starter. You will be requested to depress the start push button and observe the amount of voltage at the motor starter coil.

   a. What is the amount of voltage that is measured at the coil when the motor starter is energized? _________

   b. What is the amount of current flowing to the motor when the motor starter coil is energized? _________

   c. Explain what would happen to the motor starter if the amount of current flowing through the overload heaters to the motor increased by 200% of FLA.

   ____

   ____

   d. Identify the reset button for the motor starter. In the event of an overcurrent, the overload will trip on the motor starter, and you will have to depress the reset button for the overload to get the motor starter to energize again. Explain where the reset button is. _________

Checking Out

When you have completed this lab exercise, clean up your area, return all tools and supplies to their proper place, and check out with your instructor. Your instructor will initial here to indicate you are ready to check out. _______

CHAPTER 13

Thermostats, Pressure Controls, and Timers

OBJECTIVES

At the end of this lab exercise you will be able to:

1. Explain the operation of the heating thermostat.
2. Explain the operation of the cooling thermostat.
3. Explain the operation of the fan switch on the thermostat.
4. Understand the difference between a low-voltage and line-voltage thermostat.

INTRODUCTION AND OVERVIEW

When you are making a troubleshooting call as an HVAC technician, you will have to test a variety of thermostats. This chapter will provide exercises for you to operate these components and understand their operation. You will learn about the heating thermostat and the cooling thermostat for HVAC systems, as well as the line-voltage thermostat. Some thermostats are designed as heating thermostats, while others are designed for controlling cooling systems. You may also encounter a heat/cool thermostat that is designed to control both heating and cooling equipment. All thermostats have an element that senses temperature and causes their contacts to change state. The thermostats have developed a standard for identifying terminals. In the cooling thermostat, the contacts between the R and Y terminals are used to control the air-conditioning equipment, and they will close when the room temperature becomes warmer than the thermostat setpoint. The heating thermostat uses the contact between R and W to control the heating equipment and they will close when the room temperature goes below the setpoint. The fan switch is located between the R and G terminals and has two settings: auto and on. In the on position, the thermostat sends a constant signal to terminal G all the time, which causes the fan relay to energize and run the furnace fan. When the fan switch is in the auto position, terminal G gets a signal only when the cooling system is energized. The furnace fan is also the evaporator fan in the split-system air conditioner. In this exercise, you will check out a heating and cooling thermostat. You will learn to install and troubleshoot the thermostat and to test that it is operating correctly.

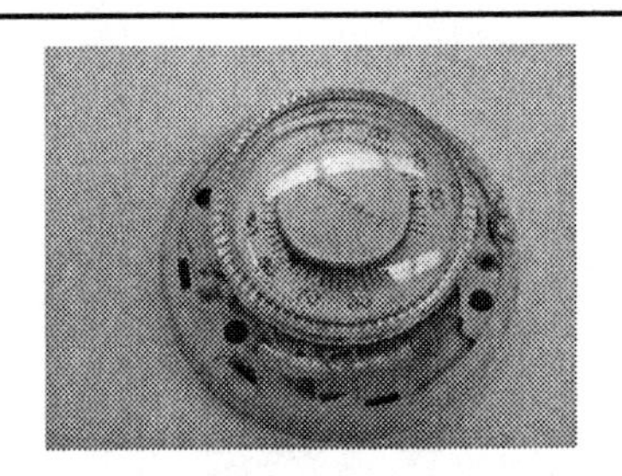

Figure 13–1 Typical mercury bulb-type thermometer. *(Courtesy of Honeywell)*

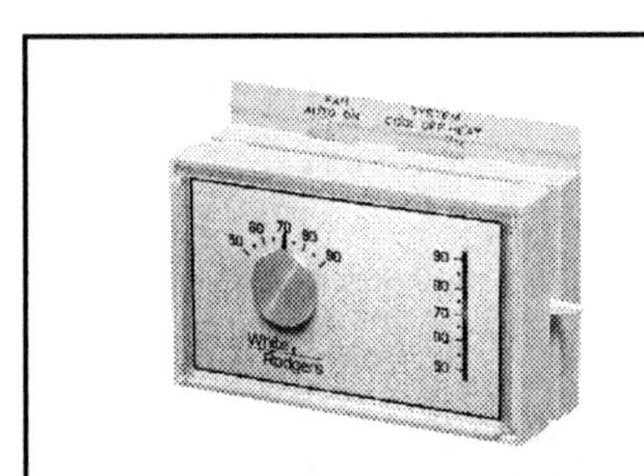

Figure 13–2 Typical horizontal mounted thermostat. *(Courtesy of White-Rodgers)*

TERMS

Airflow switch
Bimetal element
CAD cell
Cooling thermostat
Cut-in pressure
Cut-out pressure
Defrost timers
Direct-spark ignition
Dual pressure switch
Fan and limit control
Fan switch
Fixed-pressure switch
Flame rod
Gas furnace control
Gas valve
Heat anticipator
Heating thermostat
High-pressure switch
Hot surface igniter
Humidistat
Limit switch
Line-voltage thermostat
Low-pressure switch
Oil burner control
Oil pressure switch
Pilot
Primary control
Programmable thermostat
RTD (resistive temperature detectors)
Short-cycle protection
Thermistor
Thermocouple
Thermostat
Thermostat subbase
Timer control

MATCHING

Place the letter A–V for the definition from the list that matches with the terms that are numbered 1–22.

Score __________

1. ____ Fan switch
2. ____ Oil pressure switch
3. ____ Limit switch
4. ____ Airflow switch
5. ____ Cooling thermostat
6. ____ Cut-in pressure
7. ____ Cut-out pressure
8. ____ Defrost timers
9. ____ Dual pressure switch
10. ____ Fixed-pressure switch
11. ____ Heat anticipator
12. ____ Heating thermostat
13. ____ High-pressure switch
14. ____ Humidistat
15. ____ Line-voltage thermostat
16. ____ Low-pressure switch
17. ____ Pilot
18. ____ Primary control
19. ____ Programmable thermostat
20. ____ Short-cycle protection
21. ____ Thermostat subbase
22. ____ Timer control

A. The part of the thermostat (temperature control switch) that controls the cooling cycle of the HVAC system.
B. A switch that is used in conjunction with an HVAC or refrigeration compressor to ensure that the proper amount of oil remains in the compressor and does not get pumped away from it.
C. The pressure in a refrigeration or HVAC system that is set to cause the contacts of a pressure switch to open.

D. The switch on the thermostat that has two positions: The first position (on) is used to turn the furnace fan on constantly, and the other position (auto) connects the fan relay to the temperature activation part of the thermostat so that the fan cycles on and off with the air-conditioning compressor.
E. A thermostat that controls the heating equipment, which can be electrical resistance heat, gas heat, oil heat, hot water heat, steam heat, heat pump, or other type of heating system. The typical terminals on the heating thermostat are R and W.
F. A variable resistor found in a heating thermostat. The variable resistor that is used as a heat anticipator is connected in series with the gas valve, and it is set to match the amount of current that is drawn through it.
G. An electrical switch that is activated by motion. The switch senses linear or rotational motion, which causes its contacts to change state at some specified location.
H. The pressure in a refrigeration or HVAC system that is set to cause the contacts of the pressure switch to close.
I. A timer specifically designed to energize a defrost cycle and terminate it after a specified time.
J. A special control that integrates an electronic or motor-driven timer with multiple sets of contacts.
K. A control device for HVAC and refrigeration systems that provides a time delay that is connected in series with the compressor contactor coil.
L. A pressure switch that has normally closed contacts that are opened when pressure decreases below a predetermined setting.
M. A switch that is mounted in the ductwork in an air-conditioning or heating system. The switch is activated by air flow and usually has a piece of metal (a paddle) that is mounted on the switch actuator and is placed in the flow of air.
N. A pressure switch that has normally closed contacts that are opened when pressure increases above a predetermined setting.
O. A base where the thermostat is mounted.
P. The small flame for a gas furnace that is used to ignite the main burner flame.
Q. A thermostat that senses the room temperature and controls the HVAC system according to a schedule established by the homeowner. The thermostat contains an electronic chip that accepts a schedule to turn on or off heating and cooling functions.
R. A control that senses the amount of humidity in the air.
S. The part of the oil burner system that provides a high voltage spark that is used to ignite the oil during the ignition process.
T. A pressure switch that has a fixed setpoint.
U. A thermostat whose contacts are rated for 120 VAC or 240 VAC.
V. A switch that has two pressure chambers that can sense two separate pressures, such as the oil pressure and refrigerant pressure.

TRUE OR FALSE

Place a *T* or *F* in the blank to indicate if the statement is true or false.

Score ___________

1. ____ The contacts of the heating thermostat go closed when temperature falls below the setpoint.

2. ____ The gas valve is basically a solenoid valve with a safety circuit to ensure that the pilot light is burning.

3. ____ The oil burner control uses 24 V to provide the spark to the ignition points.

4. ____ The fan and limit switch provides an operational control to energize the fan, and a safety circuit to de-energize the gas valve if the temperature gets too high.

5. ____ The heat anticipator produces heat, while the heating thermostat contacts are in the open position.

6. ____ The oil pressure switch requires a time delay to allow the oil pressure to build up when the compressor is first turned on.

7. ____ A fixed-pressure switch can be adjusted a small amount around a fixed-pressure setpoint.

8. ____ The cut-in and cut-out pressure settings on an oil pressure switch will be the same pressure.

9. ____ A defrost timer is normally used to stop the compressor periodically and energize a defrost heating element.

10. ____ A commercial timer can be used to cycle off large commercial air-conditioning systems on weekends and other times when no one is in a building.

MULTIPLE CHOICE

Circle the letter that represents the correct answer to each question.

Score __________

1. The fan switch in the thermostat has:
   a. two positions (auto and on).
   b. three positions (auto, off, on).
   c. two positions (off and on).

2. The gas valve has a button that must be depressed when the pilot light is being lighted to:
   a. bypass the high-temperature safety circuit.
   b. bypass the low-temperature safety circuit.
   c. overcome the pull-in current of the solenoid in the valve until the thermocouple can produce sufficient current to hold in the solenoid.

3. The cooling anticipator in the cooling thermostat produces:
   a. heat when the contacts of the cooling thermostat are open.
   b. heat when the contacts of the cooling thermostat are closed.
   c. a cooling effect when the contacts of the cooling thermostat are open.

4. The direct-spark ignition system provides a spark to:
   a. start the pilot flame.
   b. start the main flame.
   c. to the thermocouple.

5. The flame rod in the direct-spark ignition system detects:
   a. main flame and pilot flame.
   b. only the pilot flame.
   c. only the main flame.

6. An oil pressure switch compares oil pressure with:
   a. high pressure.
   b. low pressure.
   c. a setpoint of 15 psi.

7. An oil pressure switch needs:
   a. a small amount of time delay to allow the compressor to build up oil pressure.
   b. a small amount of time delay to allow the compressor to build up its high pressure.
   c. no time delay because the oil pressure must be detected immediately or the switch will shut the compressor off so that no damage occurs if the oil pressure is low on startup.

8. A low-pressure switch connects the compressor or the coil of the compressor contactor to terminal:
   a. 4, since it is part of the normally closed contacts.
   b. 4, since it is part of the normally open contacts.
   c. 3, since it is for the time delay.

9. When you are troubleshooting an air-conditioning system and find that it is shut down because its low-pressure switch is open, you should:
   a. test the switch for continuity a second time, and replace it if it is open.
   b. test the switch for continuity a second time, and replace it if it is closed.
   c. look for the cause of the low pressure.

10. If a defrost timer does not cycle the system through a defrost cycle periodically, you should suspect that the:
    a. timer motor is not running continuously, or is not running at all.
    b. defrost heater is bad.
    c. compressor cannot shut off.

LAB EXERCISE: CHAPTER 13–1 OBSERVE AND UNDERSTAND THERMOSTATS

Safety for this Lab Exercise

In this lab exercise you have several thermostats that you will install, operate, and troubleshoot. Your instructor will provide a thermostat that controls a heating system, one that controls a cooling system, and a line-voltage thermostat. The low-voltage thermostat system operates on 24 VAC and the line-voltage thermostat works on 120 VAC or 240 VAC. You must be aware of the safety problems when a heating system or cooling system are running, and of exposed electrical terminals. You will also need to be aware of motors that are turning a fan or other mechanical device such as belts and pulleys if they are used, and to stay clear of all moving parts.

Tools and Materials Needed to Complete the Lab Exercise

Your instructor will provide a number of low-voltage and line-voltage thermostats that are connected to operating HVAC systems. You will have a heating thermostat connected to a gas heating system, a cooling thermostat that is connected to a packaged air conditioner, and a heat/cool thermostat that is connected to a split-system air conditioner with a gas furnace. You will also have a line-voltage thermostat that is connected to electric baseboard heat. You will have one station set up in this lab where you will be requested to install a thermostat subbase and connect the wires to the split-system equipment.

References to the Text

Refer to Chapter 13 in the textbook for additional information. You may need to read sections of the chapter again to help you understand the material in this exercise.

Sequence to Complete the Lab Task

Identifying Terminals on a Heat/Cool Thermostat

Figure 13–3 shows a diagram of a heat/cool thermostat. You can see the terminals on the thermostat in the diagram are identified. The terminals on thermostats have become standardized so that if you encounter a new thermostat, the terminal connections will be identified with the same letters as all other thermostats. Figure 13–4 shows the electrical symbol for the heating thermostat, and Figure 13–5 shows the electrical symbol for the cooling thermostat. Refer to the diagram in Figure 13–3 and answer the following questions.

1. What letters are used to identify the terminals of the low-voltage transformer? __________
2. What is the letter on the thermostat used to identify the "hot" terminal? __________
3. What is the letter used to identify the cooling circuit? __________
4. What is the letter used to identify the heating circuit? __________

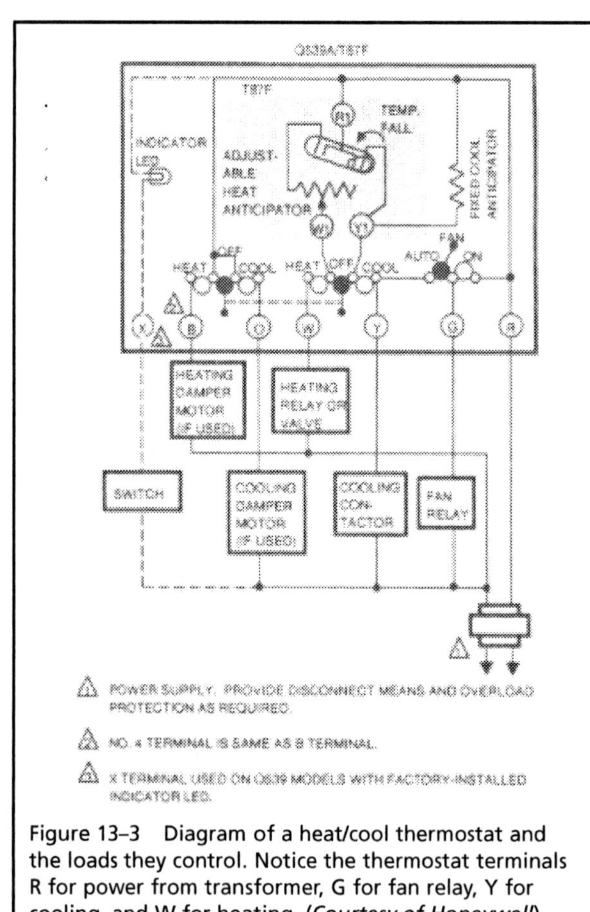

Figure 13–3 Diagram of a heat/cool thermostat and the loads they control. Notice the thermostat terminals R for power from transformer, G for fan relay, Y for cooling, and W for heating. (*Courtesy of Honeywell*)

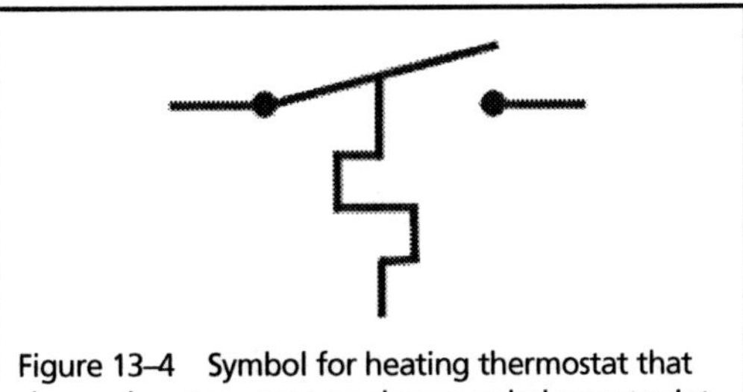

Figure 13–4 Symbol for heating thermostat that closes when temperature decreases below setpoint.

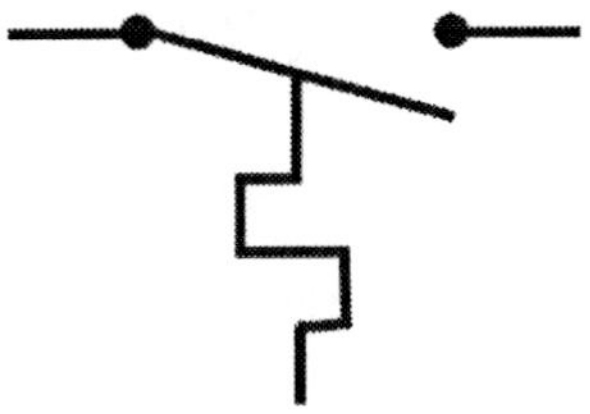

Figure 13–5 Symbol for cooling thermostat that closes when temperature increases above setpoint.

5. What do the letters R, W, Y, and G stand for?

____

____

Sequence to Complete the Lab Task

Installing a Subbase for a Heat/Cool Thermostat

1. Your instructor will provide a low-voltage heat/cool thermostat with its subbase for you to install. Be sure the power to your system is turned off while you install the thermostat and interconnecting wires. You will mount the subbase on the wall where the instructor indicates and you should ensure that it is mounted level. Next, you will run the wires and correctly connect the colored wires to the proper terminals on the thermostat subbase and at the split-system air-conditioning system that has a gas furnace. After you have the subbase correctly wired, your instructor will inspect your work and indicate it is correct. You can mount the thermostat on the subbase and test it out. Follow the steps below to mount the thermostat and test it.

2. Turn power on to the system and set the thermostat for heating. Set the thermostat setpoint for the highest temperature so that it ensures the system is "calling for heating."

3. Did the furnace come on and ignite? __________

4. Turn power on to the system and set the thermostat for cooling. Set the thermostat setpoint for the lowest temperature so that it ensures the system is "calling for cooling."

5. Did the air-conditioning compressor and condenser fan come on? __________

6. Did the evaporator fan come on? __________

7. Set the thermostat setpoint for the highest temperature setting to turn off the air-conditioning system. Did the air-conditioning system turn off? __________

8. Did the evaporator fan turn off? __________

9. Now set the fan switch on the thermostat to the on position. Did the evaporator fan turn on? __________

10. Have your instructor check your system to verify that you have wired the thermostat correctly.

11. Leave this system connected for the troubleshooting portion of this exercise.

Sequence to Complete the Lab Task

Troubleshooting the Low-Voltage Thermostat

1. In the previous exercise, you installed a low-voltage heat/cool thermostat to a heating and split-system air-conditioning system. You will use the low-voltage thermostat to operate this system and your instructor will put problems into the system for you to troubleshoot. Your instructor should supervise this section of the lab exercise. Have your instructor initial here to indicate they are present during this part of the lab. _______

2. Turn power on to the system and set the thermostat for heating. Set the thermostat setpoint for the highest temperature so that it ensures the system is "calling for heating."

3. Did the furnace come on and ignite? __________

4. Measure the voltage at terminal R to the terminal C on the furnace. Record this amount of voltage. __________

5. If you have 24 volts AC at terminal R to terminal C on the furnace, it indicates the transformer is operating correctly and is supplying low voltage to the thermostat and controls. If you do not have 24 volts AC at these two terminals, you should verify that your transformer primary winding is powered with 110 VAC. Is your transformer working correctly? __________

6. Measure the voltage at terminal W to C at the furnace. Record the voltage. __________

7. If you have 24 volts AC at terminal W to terminal C on the furnace, it indicates that the thermostat is providing voltage to the furnace and is "calling for heating." Record the voltage at terminals W to C. __________

8. Turn power on to the system and set the thermostat for cooling. Set the thermostat setpoint for the lowest temperature so that it ensures the system is "calling for cooling."

9. Did the air-conditioning compressor and condenser fan come on? __________

10. Check for 24 volts AC across terminals Y to C on the furnace. Record the voltage. __________

11. Check for 24 volts AC across terminals Y to C on the outside air-conditioning unit (condenser). Record the voltage at these terminals. __________

12. If you have 24 volts AC across terminals Y to C at the furnace, it indicates the signal is from the thermostat to the furnace. If you measure 24 volts AC at the condenser, it indicates voltage is sent to the compressor contactor coil.

13. Did the evaporator fan come on? __________

14. Measure and record the voltage at terminals G to C at the furnace. __________

15. If you have 24 volts at terminals G to C, it indicates that the thermostat is sending a signal to the fan relay in the furnace to turn on the evaporator fan.

16. Set the thermostat setpoint for the highest temperature setting to turn off the air-conditioning system. Did the air-conditioning system turn off? __________

17. Measure and record the voltage at terminals Y to C. __________

18. Did the evaporator fan turn off? __________

19. Measure and record the voltage at terminals G to C. __________

20. Now set the fan switch on the thermostat to the on position. Did the evaporator fan turn on? __________

21. Measure and record the voltage at terminals G to C. __________

22. How can you use the information about the presence of 24 volts at the terminals at the furnace when the system is calling for heating or cooling, or the furnace fan is energized?

____

____

23. If voltage is not present at the furnace terminals W to C when you want the furnace to come on, what would you suspect is the problem?

____

____

24. If voltage is not present at the furnace terminals Y to C when you want the air conditioner to come on, what would you suspect is the problem?

____

____

25. If voltage is not present at the furnace terminals G to C when you want the furnace fan to come on, what would you suspect is the problem?

____

____

26. Have your instructor put a problem in your thermostat system that allows you to troubleshoot and take voltage measurements to determine what the problem is. List what you think the problem is, and identify what tests you made to determine this.

____

____

Sequence to Complete the Lab Task

Installing and Troubleshooting a Line-Voltage Thermostat

1. This section of this lab exercise should be fully supervised by your instructor. Have your instructor initial here when you are ready to complete this part of the exercise. _______

2. Your instructor will provide a line-voltage thermostat that is connected to an electric baseboard heating element. Figure 13–6 shows a picture of a line-voltage thermostat and Figure 13–7 shows an electrical diagram of the line-voltage thermostat controlling a 240 VAC baseboard heater.

   Turn power to this system on and set the line-voltage thermostat to the highest temperature setting.

3. Place a clamp-on ammeter around the L1 or L2 wire. Measure and record the amount of current flowing to the electric heating element. __________

4. Turn the line-voltage thermostat to the lowest temperature setting. Measure and record the current flowing to the heating element. __________

5. Explain the operation of the line-voltage thermostat as it controls the electric heater.

   ____

   ____

LAB EXERCISE: 13–2 PROGRAMMABLE THERMOSTAT

Objectives for the Skills You Will Learn

At the end of this lab exercise you will be able to:

1. Explain the operation of the programmable thermostat.
2. Program the basic settings into a programmable thermostat.
3. Read the screen of the programmable thermostat and determine what its settings are.

INTRODUCTION AND OVERVIEW

One of the newer controls you will encounter as an HVAC technician is the programmable thermostat. The basic function of the programmable thermostat is the same as the mechanical-type thermostats, except the programmable thermostat can have multiple settings for conditions

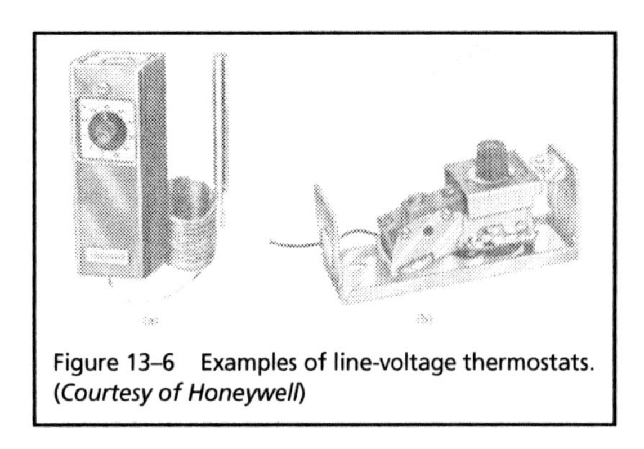

Figure 13–6 Examples of line-voltage thermostats. *(Courtesy of Honeywell)*

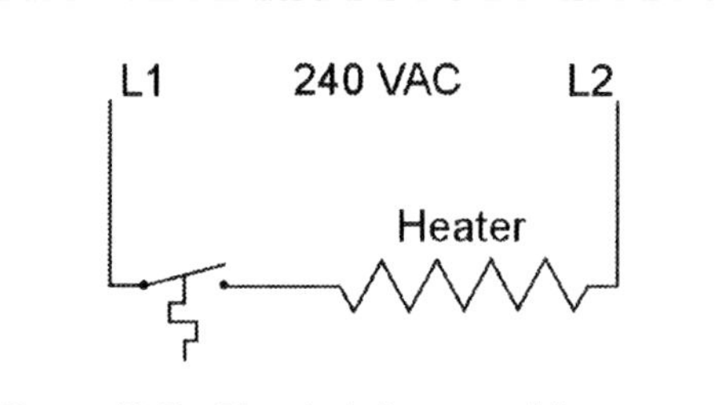

Figure 13–7 Electrical diagram of line-voltage thermostat controlling an electric heating element.

when people are sleeping or when the space is not occupied, such as when people are working if the system is in a residence, and for weekends for commercial spaces like offices.

In this exercise you will be provided a programmable thermostat and you will check its basic settings, and then make changes to the program for specific conditions. You will find that the system will need to be programmed when it is first installed, and you will need to be able to check the settings when you are called to troubleshoot a system controlled by the programmable thermostat.

Figure 13–8 shows a typical programmable thermostat. This thermostat has a touch screen, which will display a variety of setting screens that you can move through and adjust settings. Some thermostats have other methods of changing programs and settings. Figure 13–9 shows a typical screen of a touch screen-type programmable thermostat and shows where you would find the information on the thermostat. Remember the thermostat may be designed for heating only, cooling only, or it may be a heat/cool thermostat that can control both the heating equipment and the cooling equipment. If the thermostat is a heating/cooling thermostat, it will have a switch to set the system from heating or cooling. It also has a way to adjust the setpoint on the thermostat to adjust the point where the thermostat will energize. The thermostat will also have a way to indicate the current temperature of the space where it is mounted. All thermostats will have a fan switch that allows the thermostat to control the evaporator fan so that it will run all the time in the on position, and it will run with the compressor when it is in the auto position.

Figure 13–10 shows the programming and configuration pages for the White-Rodgers programmable thermostat. You can see the different parameters that you can change in the thermostat. You can use this page as an example of what can be changed on your thermostat program. Have your instructor provide the programming information for your thermostat or go on the Internet and find the information you need. When you have the information for programming, you can continue to the next steps of this exercise.

Safety for this Lab Exercise

In this lab exercise you will be provided a programmable thermostat that is connected to voltage and will control an HVAC system. You will be requested to take data from the screens of the thermostat and identify what the program is. You should be aware that the thermostat is controlling 24 VAC, and that the HVAC system that it is controlling will turn on and off when the temperature changes. You should stay clear of any moving parts in the system such as fans, and stay clear of any electrical connections.

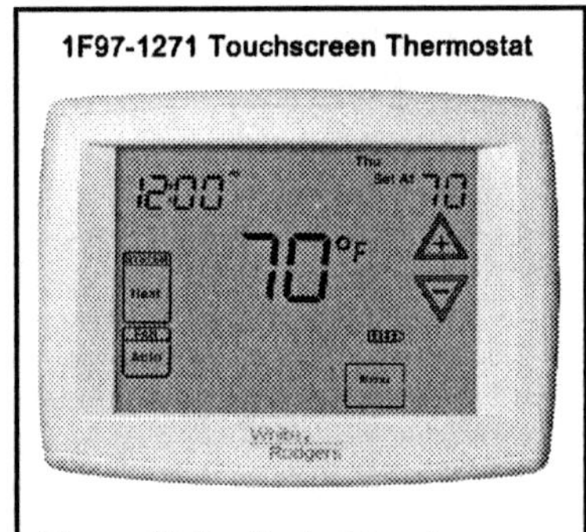

Figure 13–8 Typical touch screen programmable thermostat. (*Courtesy of White-Rodgers*)

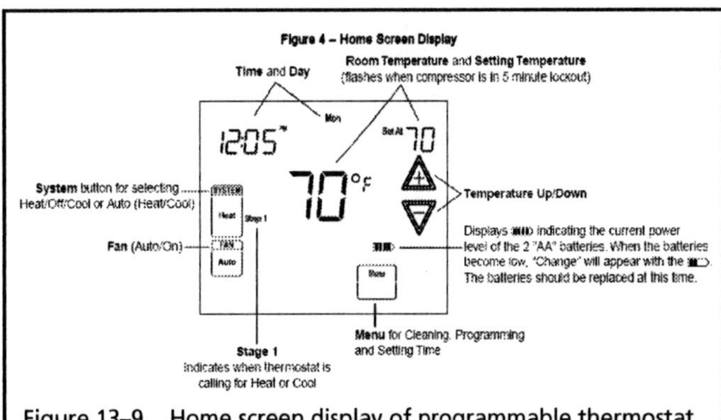

Figure 13–9 Home screen display of programmable thermostat. (*Courtesy of White-Rodgers*)

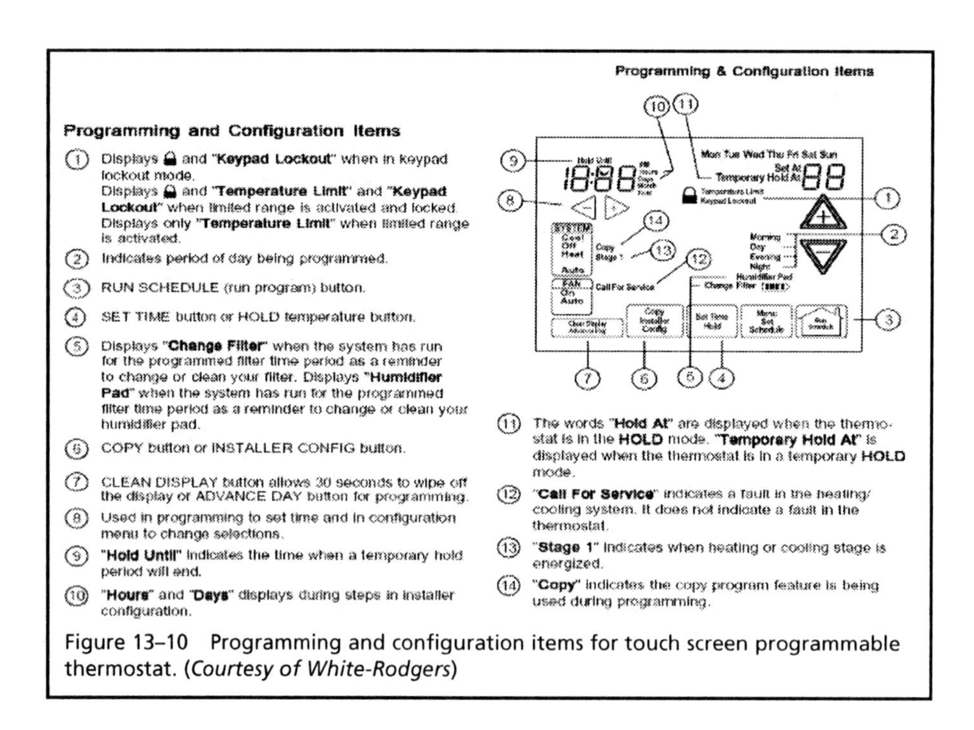

Figure 13–10 Programming and configuration items for touch screen programmable thermostat. (*Courtesy of White-Rodgers*)

Tools and Materials Needed to Complete the Lab Exercise

Your instructor will provide a programmable thermostat that is connected to an HVAC system. You will be requested to observe the program and settings in the thermostat and record this information. You will put settings into the thermostat and observe the system as the temperature changes around the thermostat.

References to the Text

Refer to Chapter 13 in the textbook for additional information. You may need to read sections of the chapter again to help you understand the material in this exercise.

Sequence to Complete the Lab Task

Observing the Program and Settings in a Programmable Thermostat

1. Your instructor will provide a programmable thermostat that is controlling an HVAC system. The programmable thermostat is powered with a battery so you will be able to read its settings even though the power is turned off to the system. Make sure the power to the system is turned off at the disconnect switch so you can make changes to the settings in the thermostat and observe the thermostat screens. Have your instructor check your information for accuracy.

   a. What is the current temperature of the room? __________

   b. What is the setpoint for the cooling system? __________

2. Sketch the face or screen of your thermostat and identify where you would find the current temperature and the setpoint.

3. Identify how you would turn on the fan on your thermostat.

____

____

4. If your thermostat is a heat/cool thermostat, identify the switch that changes the system from heating to cooling. __________

5. Set your thermostat system to cooling cycle and change the setpoint for cooling to the highest value it can accept. This should keep the compressor from turning on when you apply power to your unit. Turn the fan switch to the auto position. Next, turn power to your system on at the disconnect switch. Is the evaporator fan running? __________

6. Turn the fan switch to the on position. Does the evaporator fan for your system turn on? __________

7. Turn the fan switch to the auto position. What happens to the evaporator fan? Why?

____

____

8. Set the temperature setting for the system to the lowest value that it will accept. What happens to the compressor when the setpoint on the thermostat goes below the temperature in your room?

____

____

9. When will the compressor turn off?

____

____

10. Adjust the setpoint upward slowly until the compressor turns off. When will the compressor turn on again?

____

____

11. Explain why you must be able to read the program in a thermostat if you are on a troubleshooting call and the heating or cooling systems will not turn on.

____

____

Sequence to Complete the Lab Task

Setting the Program of Your Thermostat

1. Since your thermostat is programmable, it has a way to set different setpoint temperatures whether the space the thermostat is controlling is occupied or not. Most thermostats will have a clock/calendar that is backed up by a battery. You will need to set the current day

of the week and time of day for the thermostat to operate correctly. For example, if your thermostat is used to control a residential space in a home, the homeowner may want to set the heating thermostat so that the setpoint during the night from 11 P.M. on is set for a cooler temperature such as 65°. The thermostat is programmed to use the setpoint of 70° at 7 A.M. This feature may be called night setback. The program in some thermostats has the ability to set the program for weekdays and for weekends. These thermostats will also have a way to create a program of settings for the air-conditioning system as well. Other, more complex thermostats may allow for multiple setpoints for each day. Identify they way your thermostat is programmed.

____

____

2. How many days of the week can your thermostat be programmed for? __________

3. If your thermostat has the ability to identify the time of day and the days of the week, adjust your thermostat for the correct time and day of the week you are doing this lab. Have your instructor check your settings. __________

4. How may different programs will your thermostat accept? __________

5. Create a program and enter it into your thermostat that uses the setpoint of 70° for heating on Monday through Friday, from 7 A.M. to 11 P.M. Use the setpoint of 65° for every day of the week for night setback from 11 P.M. to 7A.M. Have your instructor check your settings when you have them entered.

6. If you have time and your instructor will allow it, create a program that changes the setpoint temperature settings for the times and days you are in the lab at your class. For this exercise, you will need to return to the thermostat several times during the week at different times and check to see if its program is making the changes you have programmed in. If you do not have this capability in your lab, you may skip this step.

7. Some thermostats have a default program that provides original factory settings for your thermostat. This allows you to set the thermostat back to the default settings if the program does not seem to operate correctly. Sometimes settings can be put into the thermostat that conflict with each other and will prevent the system from turning on the heating or cooling system. Does your thermostat have a pre-program or a default program that comes with factory settings? __________

8. If your thermostat has a default program with factory settings, read the instructions and identify how to reset the thermostat to the factory default settings. Explain how to set the thermostat default settings.

____

____

9. Set your thermostat to the default settings and call your instructor to check your work to ensure the program returned to the original settings.

Sequence to Complete the Lab Task

Changing the Battery on the Programmable Thermostat

1. Programmable thermostats use a battery as backup power to keep the program in the system in case the electrical power gets turned off. The thermostat will have an indicator to let you know that the battery needs to be changed. It is important that you keep the electrical power on while you change the battery. If you turn off power to the system and remove the battery, all of the settings in the programmable thermostat will be lost. Your instructor will provide a thermostat that has a low battery that you can use for this exercise. Leave the power to the system on and locate the battery in the thermostat. Change the battery and check the indicator lamp to ensure it goes out and the system returns to its normal state. Have your instructor check your work.

2. Did the thermostat keep its settings after you changed the battery? __________

3. Explain why it is important to keep power on to the system when you change the battery.

   ____

   ____

LAB EXERCISE: CHAPTER 13–3 PRESSURE CONTROLS

Objectives for the Skills You Will Learn

At the end of this lab exercise you will be able to:

1. Explain the operation of a high-pressure switch.
2. Explain the operation of a low-pressure switch.
3. Explain the operation of the oil pressure switch.
4. Explain how a pressure switch can be used as a temperature control.

INTRODUCTION AND OVERVIEW

Pressure switches are used to protect the HVAC and refrigeration systems against low-pressure or high-pressure conditions. The low-pressure switch may also be used as a temperature control similar to a thermostat since refrigeration pressure has an equivalent temperature rating. Figure 13–11 shows an example of a low-pressure switch and its electrical symbol, and Figure 13–12 shows an example of a high-pressure switch and its electrical symbol. The high-pressure switch is usually connected in series with the coil of the compressor contactor and its contacts will open and de-energize the coil of the compressor contactor any time the refrigeration system

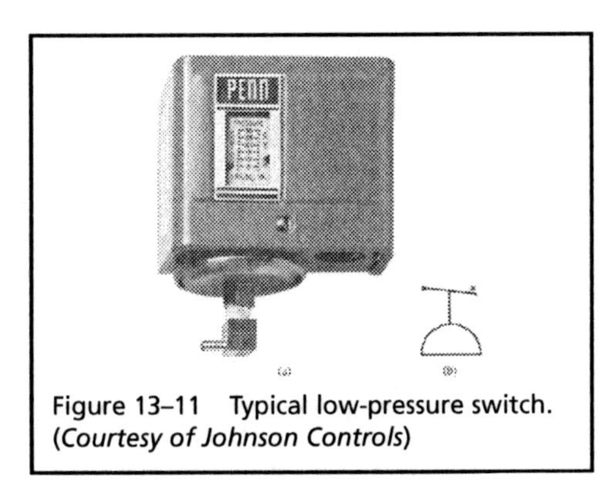

Figure 13–11 Typical low-pressure switch. *(Courtesy of Johnson Controls)*

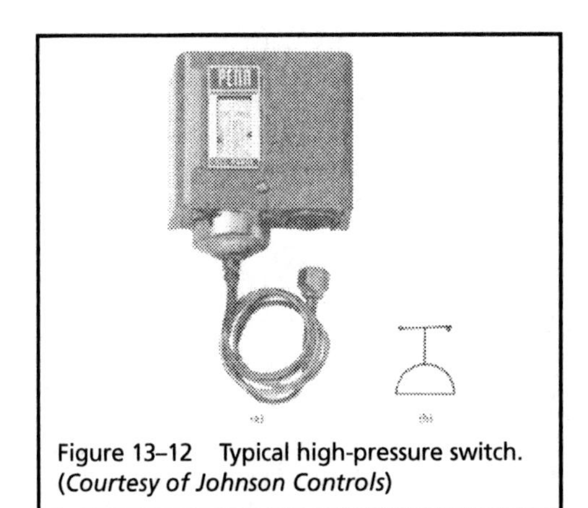

Figure 13–12 Typical high-pressure switch. *(Courtesy of Johnson Controls)*

pressure increases above the preset pressure. This condition can occur if the condenser coils get dirty, or if the condenser fan stops moving air across the condenser coils. If the pressure gets too high, it can cause the compressor to draw too much current and overheat and possibly do damage to the compressor motor. When the high-pressure switch trips, it will remain in the open position until you manually reset it. It is important to check for the problems that can cause the high-pressure condition, and correct it before you reset the switch.

The low-pressure switch is also used to protect the compressor. It is usually connected in series with the coil of the compressor contactor. If the system has low refrigeration pressure, the low-pressure switch will open and de-energize the coil to the compressor contactor. Conditions that can cause low pressure include a leak in the refrigeration piping, low air movement across the evaporator coil due to dirty filters, poor duct work, registers being closed, or loose belts on the evaporator fan if it is belt driven. It is important that you check for the problems that cause the low pressure before you reset the low-pressure switch. In this lab you will be requested to check the low-pressure and high-pressure switches and observe their operation.

Safety for this Lab Exercise

In this lab exercise you have several pressure switches that you will observe their operation and troubleshoot them. Your instructor will provide a system with a high-pressure switch and a low-pressure switch. In some applications the pressure switches will be connected in a low-voltage system, and in some refrigeration systems the controls will be wired in the line-voltage circuit. You will need to be aware of the problems that high refrigeration pressure or low refrigeration pressure may cause. You should also be aware that when an HVAC or refrigeration system experiences higher pressures, the temperature in the refrigeration tubing will become higher and cause the possibility of burning the skin if you come into contact with the refrigeration line. You must also be aware of any exposed electrical connections when you are checking the switches or making any voltage measurements.

Tools and Materials Needed to Complete the Lab Exercise

Your instructor will provide a low-pressure switch and a high-pressure switch that are not connected to a system, and you can check them out on a bench and observe their data plates. You will also be provided a low-pressure switch and high-pressure switch that are connected to an operating HVAC or refrigeration system. You will need a voltmeter and refrigeration gauges for this exercise. Your instructor can install the refrigeration gauges on the system so you can observe the pressure when the switches open and close.

References to the Text

Refer to Chapter 13 in the textbook for additional information. You may need to read sections of the chapter again to help you understand the material in this exercise.

Sequence to Complete the Lab Task

Identifying Terminals and Data Plate on a High-Pressure Switch and Low-Pressure Switch

Your instructor will provide a high-pressure switch on a bench for you to check out and take data from. Answer the following questions in regard to your high-pressure switch.

1. What is the pressure setting for your high-pressure switch? __________

2. What is the voltage rating of the switch contacts in your high-pressure control? __________

3. What is the current rating of the switch contacts in your high-pressure control? __________

4. Test the contacts of your high-pressure switch with an ohmmeter. Are the contacts open or closed? __________

5. When will the contacts in your high-pressure switch change state? __________

6. When you are on a troubleshooting call and you find the high-pressure switch tripped, what should you check before you complete the call? __________

7. Name two conditions that can cause high pressure in an HVAC or refrigeration system.

   ____

   ____

8. Your instructor will provide a low-pressure switch on a bench for you to check out and take data from. Answer the following questions in regard to your high-pressure switch.

   a. What is the pressure setting for your low-pressure switch? __________

   b. What is the voltage rating of the switch contacts in your low-pressure control? __________

   c. What is the current rating of the switch contacts in your low-pressure control? __________

   d. Test the contacts of your low-pressure switch with an ohmmeter. Are the contacts open or closed? __________

   e. When will the contacts in your low-pressure switch change state? __________

   f. When you are on a troubleshooting call and you find the low-pressure switch tripped, what should you check before you complete the call? __________

   g. Name two conditions that can cause low pressure in an HVAC or refrigeration system.

   ____

   ____

Sequence to Complete the Lab Task

Observing the Operation of the High-Pressure Switch and Low-Pressure Switch and Steps for Troubleshooting Them

Your instructor will provide a high-pressure switch and low-pressure switch that are connected to an HVAC or refrigeration system for you to observe their operation. Answer the following questions in regard to the high-pressure and low-pressure switches.

1. Where is the refrigeration line of the high-pressure switch connected to? __________

2. Where is the refrigeration line of the low-pressure switch connected to? __________

3. Figure 13–13 shows a high-pressure switch in a control circuit with a low-pressure switch and thermostat controlling a compressor contactor coil. You can use this diagram, or your instructor will provide a diagram of the circuit of your system if it is different from this diagram. What are the high-pressure and low-pressure switches protecting?

   ____

4. Have your instructor turn off power to your system and trip the high-pressure switch. Use the diagram in Figure 13–14 that has numbered test points. Turn the power back on and try to run the system. Since the high-pressure switch is open, use your voltmeter to test the circuit where the high-pressure switch contacts are located. Set the voltmeter to measure 24 volts AC and place one terminal on the C terminal at test point 9. Take the other voltmeter lead and test points 1 through 7 in order and place the amount of voltage that you are measuring in the middle column of the following table. In the right column, indicate the condition of the circuit to that point. Since the high-pressure switch contacts are open, you should have voltage at test points 1, 2, and 3, and then lose voltage from test points 4 through 7. You can use this troubleshooting method to find the open in any circuit.

5. Have your instructor turn off voltage to your system and verify your readings. Discuss your conclusions and why this method of troubleshooting will work in identifying where the open is in the circuit.

Sequence to Complete the Lab Task

Testing the Oil Pressure Switch

1. Your instructor will provide an oil pressure switch for you to observe its operation and understand how it troubleshoots it. Figure 13–15 shows a picture of a typical oil pressure switch. The oil pressure switch will be connected to an HVAC or refrigeration system at a point where it can measure the oil pressure and the refrigeration pressure at the compressor. The oil pressure should be approximately 15 psi larger than the refrigeration pressure in the system. The oil pressure switch has a built-in timer to provide time for the system to build oil pressure when the compressor is first started. The oil pressure switch also has a differential pressure switch that compares the pressure in the compressor and the oil pressure. The differential between the two pressures should be approximately 15 psi. If the compressor fails to develop oil pressure within 30 seconds, the switch will trip and de-energize the compressor contactor. The problems that can cause the low oil pressure condition are a leak in the refrigeration high-pressure side, which allows oil to leak with the refrigerant, or the failure of the oil pressure switch. Very seldom is the pressure switch the problem. Rather, the oil pressure switch is checking for oil pressure and if it trips, it is simply reporting that

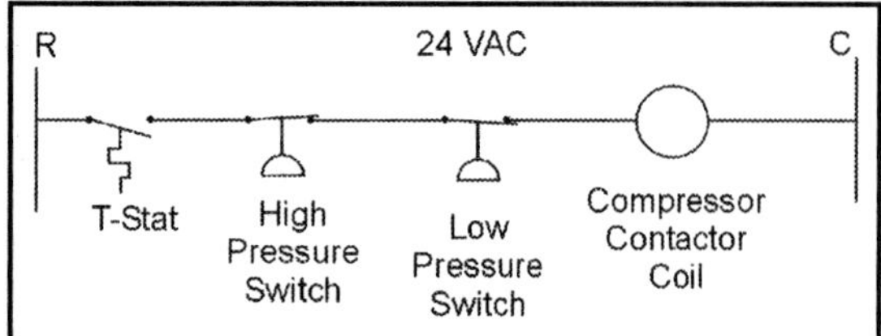

Figure 13–13 Control circuit with thermostat, high-pressure switch, low-pressure switch, and compressor contactor coil.

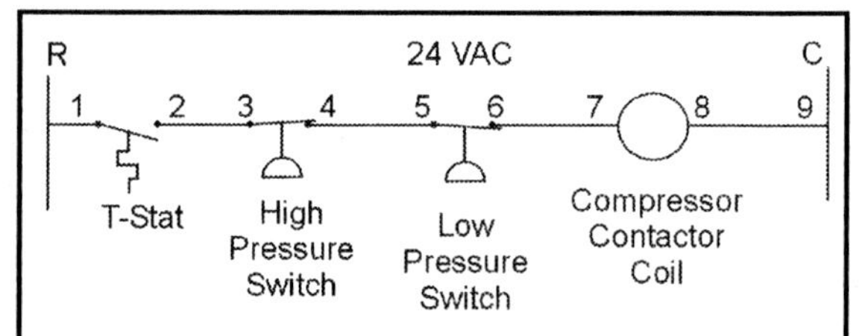

Figure 13–14 Control circuit with thermostat, high-pressure switch, low-pressure switch and compressor contactor coil. Numbers are identified in the circuit as test points for troubleshooting.

Test Point	Amount of Voltage between Test Point and Terminal C	Condition of the Circuit to this Point
1		
2		
3		
4		
5		
6		
7		

the system has an oil pressure problem. You can test to see if you are getting the correct amount of pressure differential by placing a T fitting in each of the refrigeration lines that are connected to the oil pressure switch. You can use your refrigeration gauges to read the pressure in each of these lines. If the differential between the two readings is more than 15 psi and the switch still trips, then the switch may be faulty. Generally, the differential will be less than 15 psi if the switch is tripping.

2. If the system you are troubleshooting has an oil pressure switch tripped, what would you suspect is the problem?

____

____

3. If you reset the oil pressure switch and it trips again, explain how testing the differential pressure will help you determine if the switch is bad or if the system is losing oil pressure.

____

____

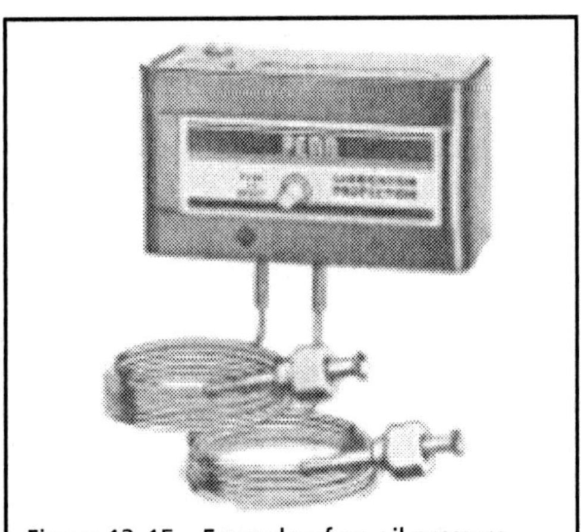

Figure 13–15 Example of an oil pressure switch. (*Courtesy of Johnson Controls*)

Checking Out

When you have completed this lab exercise, clean up your area, return all tools and supplies to their proper place, and check out with your instructor. Your instructor will initial here to indicate you are ready to check out. _______

CHAPTER 14

Controls for Gas, Electric, and Oil Heating Systems

OBJECTIVES

At the end of this lab exercise you will be able to:

1. Identify the gas valve, pilot assembly, and fan and limit switches on the standing pilot type gas furnace.
2. Identify the component parts of the high-efficiency type gas furnace with electronic ignition.
3. Identify the parts of the oil furnace.
4. Identify the parts of the electric furnace.

INTRODUCTION AND OVERVIEW

This lab exercise will show you pictures of a variety of gas, oil, and electric heating equipment. Your instructor will provide similar equipment in your lab for you to check out and identify the basic parts. You will see the basic parts and learn about troubleshooting them when you are making a troubleshooting call as an HVAC technician. You will need to recognize the parts of the newest heating systems, as well as the older type technology.

TERMS

Boiler
CAD cell
Cam stat
Direct-spark ignition
Electric furnace
Electric heating coils
Electric sequencer relay
Fan and limit switch
Fan control
Fan switch
Flame rod
Flame rollout over temperature switch
Gas valve
Heat exchanger
High-efficiency furnace
Hot surface igniter
Oil burner
Oil burner control
Protectorelay
Thermal fuse
Thermocouple

MATCHING

Place the letter A–R for the definition from the list that matches with the terms that are numbered 1–18.

Score __________

1. ____ Boiler
2. ____ CAD cell
3. ____ Cam stat
4. ____ Direct-spark ignition
5. ____ Electric furnace
6. ____ Electric heating coils
7. ____ Electric sequencer relay
8. ____ Fan control
9. ____ Flame rod
10. ____ Flame rollout over temperature switch
11. ____ Gas valve
12. ____ Heat exchanger
13. ____ High-efficiency furnace
14. ____ Hot surface igniter
15. ____ Oil burner (oil furnace)
16. ____ Oil burner control
17. ____ Thermal fuse
18. ____ Thermocouple

A. Coils of high resistance wire that produces a large amount of heat when current flows through them. The coils can be placed in a furnace and air is blown over them.
B. A cadmium sulfide cell (CdS) that changes its resistance in response to the change in light that strikes its surface. This device is primarily used in oil furnaces to detect the presence of a flame.
C. A heating system where water or other fluid is heated to a warm temperature or to a point where the fluid changes to steam. The heated fluid or steam is circulated through piping and radiators that are located in spaces where heat is needed.
D. A sensor that detects the presence of a pilot flame in the gas heating system.
E. The part of a furnace generally made of thin metal that has hot combustion gases flowing through it, and indoor air flowing over it.
F. A temperature sensor made of two dissimilar metals. When heat is applied to the tip of the thermocouple, it generates a small amount of DC millivoltage (1/1000 of a volt).
G. A furnace that burns oil as its primary fuel. The furnace consists of a burner section and a heat exchanger.
H. A specially designed fuse link that is placed in series in an electrical heating element that senses over temperature. If the temperature exceeds the setting of the fuse link, it will open and stop current from flowing through the element.
I. A gas, oil, or electrical furnace that is designed to have efficiencies 90% or above. The efficiency rating for furnaces is called Annual Fuel Utilization Efficiency (AFUE).
J. A control system that is specifically designed to protect an oil furnace. The control has sensors to determine if the fuel oil has been ignited, and it also checks against flame out.
K. A normally closed temperature sensitive switch that is precisely located to sense abnormal flame that occurs outside of the normal position where it is drawn up the chimney.
L. A special set of relays that have timers built into them, and are wired together to create a sequence that ensures one set of heater coils are energized at a time, until the second stage of electrical heating is required.

M. A brand name of a temperature sensitive switch that is used to turn on the furnace fan when the temperature increases in the plenum. It may also provide a second set of electrical normally closed contacts that are wired in series with the gas valve, and will open if the temperature in the furnace gets too high, which indicates a major problem in the furnace.
N. A switch that is mounted in a furnace that turns on the furnace fan when the temperature rises to the predetermined setpoint.
O. An electronic ignition control that creates a series of electrical pulses (sparks) across an open gap between two metal terminals. The electrical sparks are intense enough to ignite natural gas or LP gas in the pilot light.
P. The valve in a gas furnace that is controlled by a solenoid. When the solenoid is energized, the valve opens and allows full flow of gas to the burner.
Q. A furnace that uses electrical resistance coils that is its main source of heat. The electrical coils produce large amounts of heat when current flows through them.
R. A special igniter that is made of ceramic material that glows red hot when current is applied to it.

TRUE OR FALSE

Place a *T* or *F* in the blank to indicate if the statement is true or false.

Score ____________

1. ____ The hot surface igniter creates sparks that ignite the pilot flame.
2. ____ The gas valve is basically a solenoid valve with a safety circuit to ensure that the pilot light is burning.
3. ____ The oil burner control uses 24 V to provide the spark to the ignition points.
4. ____ The fan and limit switch provides an operational control to energize the fan, and a safety circuit to de-energize the gas valve if the temperature gets too high.
5. ____ The electric furnace uses an electronic type direct-spark ignition.
6. ____ The combustion blower in a high-efficiency furnace turns on after the gas valve opens and the main burner flame is established.
7. ____ The spark for the ignition system on the oil furnace turns off after the flame is established.
8. ____ After the flame is established in the gas furnace, the temperature increases until the fan switch is enabled and energizes the furnace fan.
9. ____ The pilot light must be established before the thermostat can energize the gas valve in the gas furnace.
10. ____ The differential air pressure switch in the high-efficiency gas furnace closes when the exhaust fan is moving a sufficient amount of air.

MULTIPLE CHOICE

Circle the letter that represents the correct answer to each question.

Score __________

1. If the main heating coils of an electric furnace are not heating up, you should suspect that the:
   a. coil of the fan relay is open or the fan relay contacts are faulty.
   b. fan switch contacts are not closed or the fan switch is faulty.
   c. system has an open fuse, the contacts of the first sequencer are not closed, or the main heating element is open.

2. If the main gas valve of a standing pilot gas furnace is not energized and allowing gas to flow to the main burner, you should suspect that the:
   a. heat anticipator of the thermostat is set incorrectly.
   b. pilot flame is not established, the thermocouple is faulty, or the coil in the gas valve is open.
   c. fan switch is open, the fan is faulty, or the filter may be dirty.

3. The sequence of operation of a high-efficiency furnace is:
   a. the thermostat calls for heat, the combustion fan turns on, the pilot is established, the main burner comes on.
   b. the thermostat calls for heat, the pilot is established, the combustion fan turns on, the main burner comes on.
   c. the thermostat calls for heat, the pilot is established, the main burner comes on, the combustion fan turns on.

4. The sequence of operation for an oil burner is:
   a. the thermostat calls for heat, the ignition spark is established, the furnace fan comes on, the main flame is established.
   b. the thermostat calls for heat, the ignition spark is established, the main flame is established, the furnace fan comes on.
   c. the thermostat calls for heat, the furnace fan comes on, the ignition spark is established, the main flame is established.

5. The sequence of operation of a standing pilot gas furnace is:
   a. the thermostat calls for heat, the pilot is ignited, the main gas valve is energized, the furnace fan comes on.
   b. the thermostat calls for heat, the furnace fan comes on, the pilot light is ignited, the main gas valve comes on.
   c. the pilot light is ignited and remains on continuously, the thermostat calls for heat, the main gas valve is energized, the furnace fan comes on.

6. The direct-spark ignition system provides a spark to:
   a. start the pilot flame.
   b. start the main flame.
   c. the thermocouple.

7. The flame rod in the direct-spark ignition system detects:
   a. the main flame and pilot flame.
   b. only the pilot flame.
   c. only the main flame.

LAB EXERCISE: OBSERVING AND UNDERSTANDING HEATING SYSTEMS

Safety for this Lab Exercise

In this lab exercise you will be requested to take the covers and doors off of several different types of heating systems so that you can locate and identify different parts. You must be aware that the power may be turned on to the systems so that you can observe their normal operation. You should be aware of any open electrical terminals, motors that are turning a fan or other mechanical device such as belts and pulleys if they are used, and to stay clear of all moving parts.

Tools and Materials Needed to Complete the Lab Exercise

Your instructor will provide a number of heating systems such as a standing pilot type gas furnace, a high-efficiency gas furnace with electronic ignition, an oil furnace, and an electric furnace for you to observe their operation and identify their basic parts. Your instructor may provide some of the basic parts of these furnaces for you to check out on a workbench. The parts on the workbench should not have any power applied so you can handle them safely and take a closer look at them.

References to the Text

Refer to Chapter 14 in the textbook for additional information. You may need to read sections of the chapter again to help you understand the material in this exercise.

Sequence to Complete the Lab Task

Identifying the Parts of the Standing Pilot Type Furnace

Be sure to check with your instructor for the steps where you are requested to locate parts. Indicate that you have found the location of the parts and can explain what they are and how they work.

1. Figure 14–1 shows the pilot and thermocouple of a standing pilot type furnace. Check the furnace your instructor has provided and locate the pilot and thermocouple. If the pilot light is lit, be aware that the flame must reach the thermocouple to cause it to detect the flame.

2. Figure 14–2 shows the gas valve for the standing pilot type furnace. Check the furnace your instructor has provided and locate the gas valve.

3. Figure 14–3 shows a flame rollout switch. Check the furnace your instructor has provided and locate the flame rollout switch. If your furnace does not have a flame rollout switch, identify any over temperature type switches it may have.

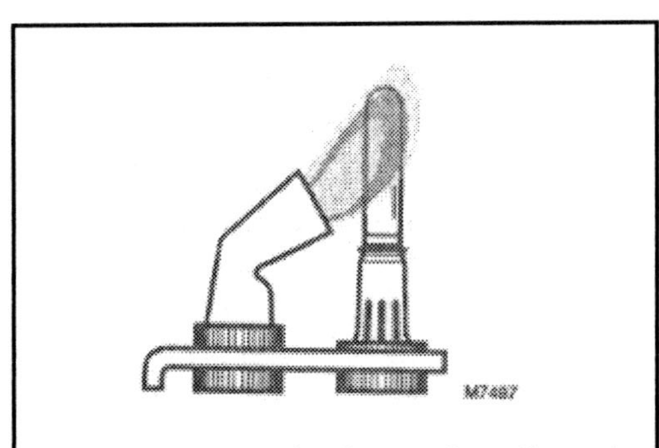

Figure 14–1 Example of a standing pilot and thermocouple. *(Courtesy of Honeywell)*

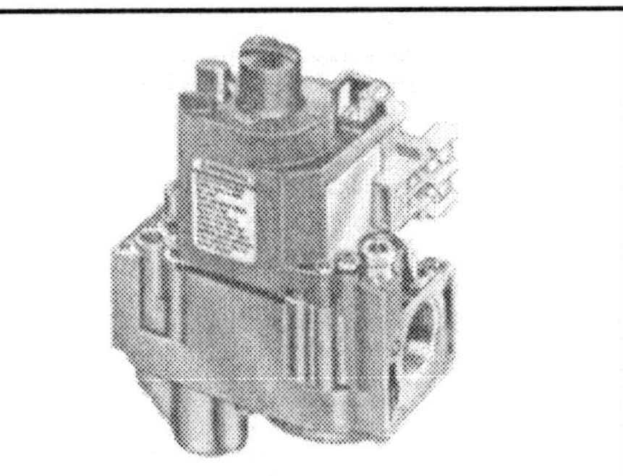

Figure 14–2 A gas valve for a standing pilot type furnace. *(Courtesy of Honeywell)*

Figure 14–3 Flame rollout switch. (*Courtesy of Therm-O-Disc*)

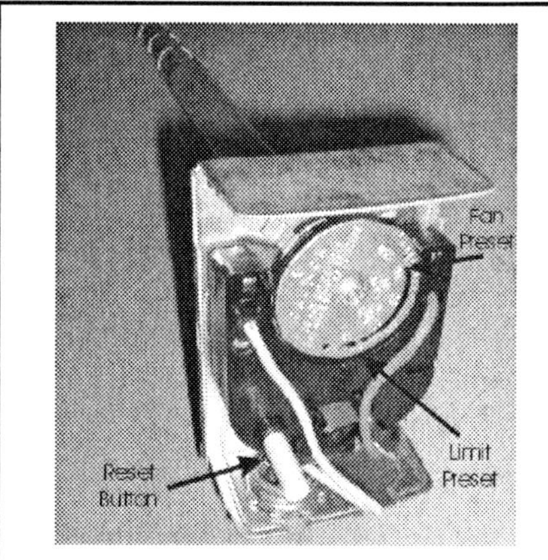

Figure 14–4 The fan and limit switch.

4. Figure 14–4 shows a fan and limit switch. Check the furnace your instructor has provided and locate the fan and limit switch.
   a. What temperature is the fan switch set for? __________
   b. What temperature is the limit switch set for? __________

Sequence to Complete the Lab Task

Identifying the Parts of a High-Efficiency Type Furnace

Be sure to check with your instructor for the steps where you are requested to locate parts. Indicate that you have found the location of the parts and can explain what they are and how they work.

1. Figure 14–5 shows the parts of a high-efficiency furnace. Check the furnace your instructor has provided and locate the following parts:

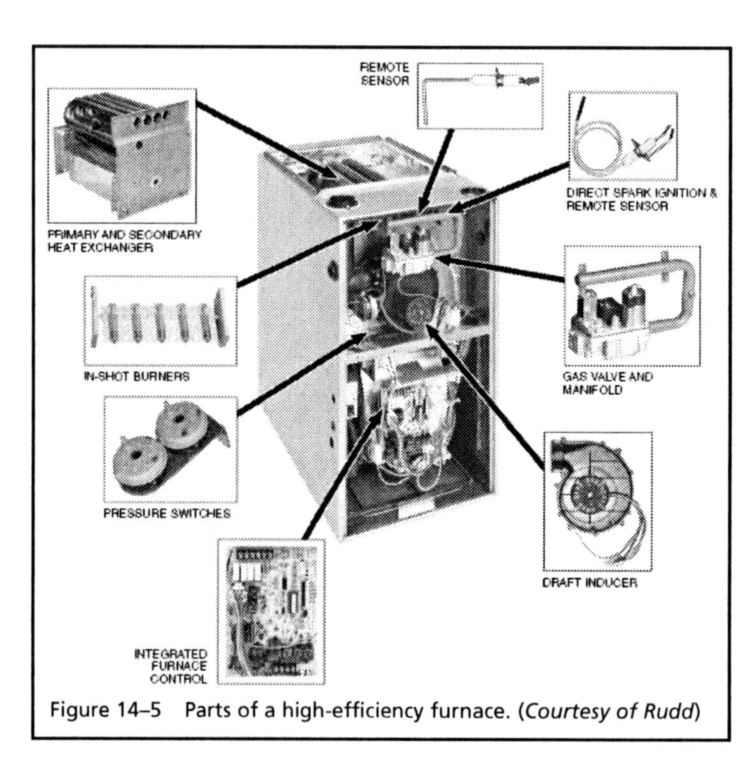

Figure 14–5 Parts of a high-efficiency furnace. (*Courtesy of Rudd*)

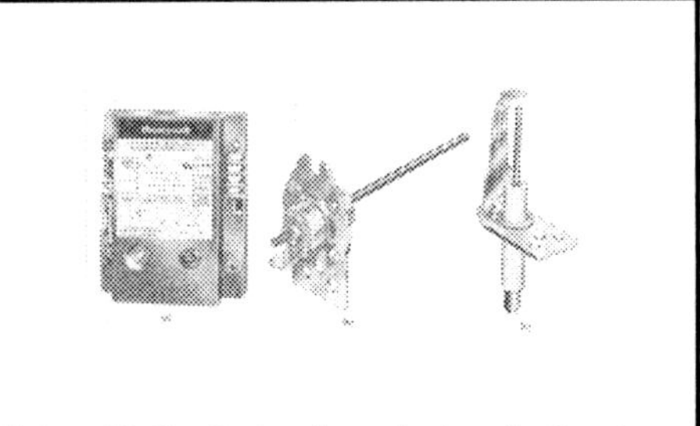

Figure 14–6 Parts of an electronic direct-spark ignition. (*Courtesy of Honeywell*)

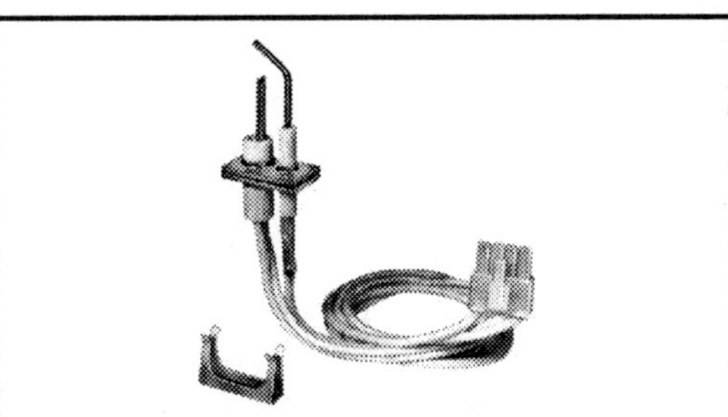

Figure 14–7 Example of a direct-spark ignition component. (*Courtesy of Honeywell*)

a. Heat exchanger
b. Gas valve
c. Direct-spark ignition or electronic ignition
d. Pressure switches
e. Draft inducer
f. The electronic control board

2. Figures 14–6 and 14–7 show examples of a direct-spark ignition system. Check the furnace your instructor has provided and locate the electronic ignition parts. If the furnace is operational, start the furnace and observe the direct-spark ignition in operation.

3. Figure 14–8 shows an example of a hot surface igniter. Check the furnace your instructor has provided and locate the hot surface igniter. If the furnace is operational, start the furnace and observe the hot surface igniter in operation.

Sequence to Complete the Lab Task

Identifying the Parts of an Oil Burner Type Furnace

Be sure to check with your instructor for the steps where you are requested to locate parts. Indicate that you have found the location of the parts and can explain what they are and how they work.

1. Figure 14–9a shows a typical oil burner assembly. Check the oil furnace your instructor has provided and locate the oil burner assembly. If your oil furnace is operational, turn on the oil burner and observe its operation.

Figure 14–8 Typical hot surface igniter. (*Courtesy of Nortonigniters*)

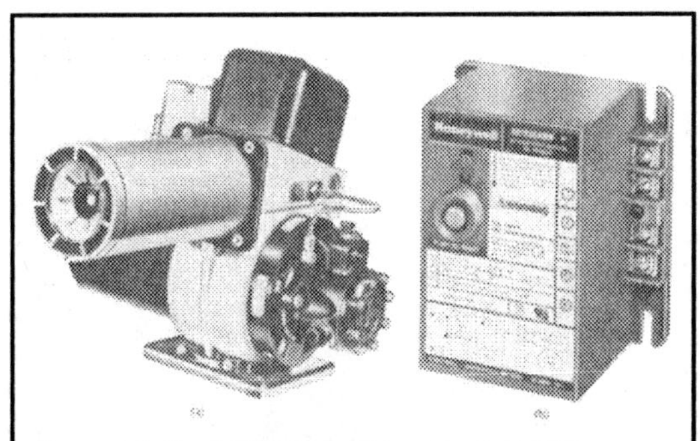

Figure 14–9 (a) Typical oil burner assembly. (*Courtesy of Beckett Corporation*) (b) Oil burner control. (*Courtesy of Honeywell*)

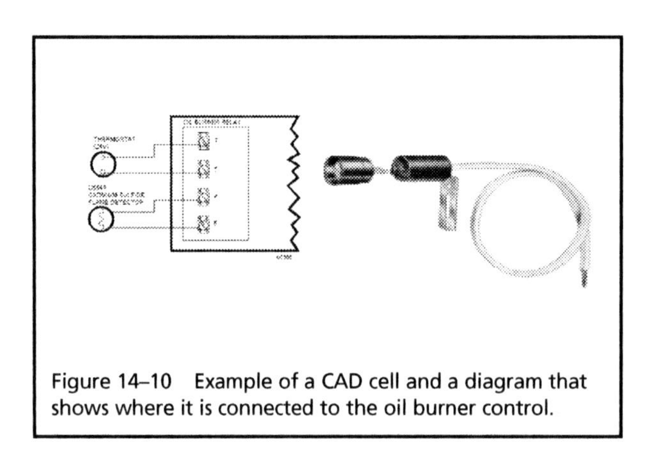

Figure 14–10 Example of a CAD cell and a diagram that shows where it is connected to the oil burner control.

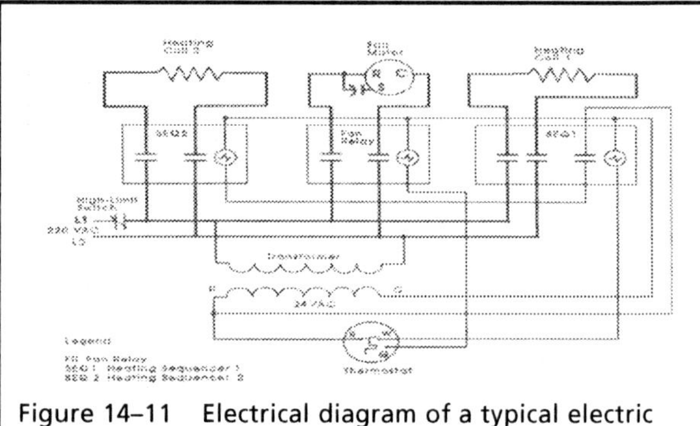

Figure 14–11 Electrical diagram of a typical electric furnace.

2. Figure 14–9b shows a typical oil burner control. Check the oil furnace your instructor has provided and locate the oil burner control.

3. Figure 14–10 shows the CAD cell for the oil burner. The CAD cell is used to detect the presence of a flame to ensure the oil burner ignites properly. Check the oil furnace your instructor has provided and locate the CAD cell on the oil burner.

Sequence to Complete the Lab Task

Identifying the Parts of an Electric Furnace

Be sure to check with your instructor for the steps where you are requested to locate parts. Indicate that you have found the location of the parts and can explain what they are and how they work.

1. Figure 14–11 shows the electrical diagram of a typical electric furnace. Check the parts on the diagram and locate them in the electric furnace your instructor has provided.

2. Figure 14–12 shows a typical electric sequencer switch for an electric furnace. Check the electric furnace your instructor has provided and locate the electric sequencer switch.

3. Figure 14–13 shows a typical electric heating coil for the electric furnace. Check the electric furnace your instructor has provided and locate the electric heating coil.

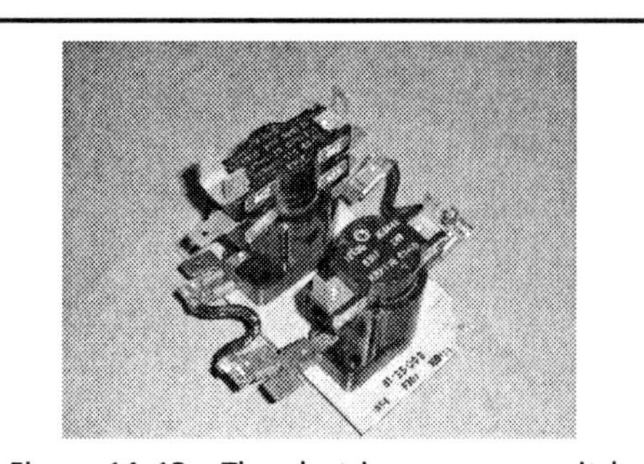

Figure 14–12 The electric sequencer switch that brings on the stage of electrical heat.

Figure 14–13 Electric heating coil for electronic furnace.

SAFETEY NOTICE: Be aware that when the electric heating coil is energized it will have voltage present on all metal surfaces of the coils which poses an electrical shock hazard. Do not allow the coils to come into contact with your skin anywhere or you can receive a severe electrical shock. Also be aware that the surface of this heating coil is extremely hot (red hot) when it is energized, and it and all surrounding areas will be very hot and can cause a severe burn if you come into contact with the hot surfaces.

Checking Out

When you have completed this lab exercise, clean up your area, return all tools and supplies to their proper place, and check out with your instructor. Your instructor will initial here to indicate you are ready to check out. _______

CHAPTER 15

Controls for Air-Conditioning Systems

OBJECTIVES

At the end of this lab exercise you will be able to:

1. Identify the basic parts of a window type air-conditioning system.
2. Identify the basic parts of a split-system type air-conditioning system.
3. Identify the basic parts of a packaged system type air-conditioning system.
4. Explain the operation of an air-conditioning system.

INTRODUCTION AND OVERVIEW

This lab exercise will show you pictures of a variety of air-conditioning equipment. Your instructor will provide similar equipment in your lab to check out and identify the basic parts. You will see the basic parts and learn about troubleshooting them when you are making a troubleshooting call as an HVAC technician. You will need to recognize the parts of the HVAC systems when you are working on them.

TERMS

Air-conditioning system
Compressor
Condenser fan motor
Evaporator fan motor
Packaged system air conditioning
Split-system air conditioner
Window unit air conditioner

MATCHING

Place the letter A–G for the definition from the list that matches with the terms that are numbered 1–7.

Score __________

1. ____ Air-conditioning system
2. ____ Compressor
3. ____ Condenser fan motor
4. ____ Evaporator fan motor
5. ____ Packaged system air conditioning
6. ____ Split-system air conditioner
7. ____ Window unit air conditioner

A. The motor for the fan that blows air over the evaporator coil. The fan also doubles as the furnace fan when the HVAC system is in the heating mode.
B. An HVAC system that has the evaporator in an indoor system such as a furnace, and the compressor and condenser is located in the outdoor unit. Refrigeration tubing connects the condenser coil, compressor, and evaporator coil.
C. A system that consists of a condensing section that is located outdoors and has a compressor and condenser coil. The indoor part of the system consists of a furnace or fan unit that

moves air over an evaporator coil. The evaporator coil is mounted in ductwork and the air moving over it cools the conditioned space.

D. An HVAC system that contains the evaporator fan, condenser fan, and compressor in one unit. This unit is specifically designed to be used as a rooftop system or a self-contained system that does not require any refrigerant piping between separate evaporator and condenser.

E. An electric motor that directly drives a piston that is used to pump refrigerant in an HVAC and refrigeration system. In a hermetically sealed compressor, the electric motor is sealed in the same chamber as the piston pump and refrigerant is used to cool this motor.

F. A small air-conditioning system that has the evaporator, condenser, and compressor all in one cabinet that fits into a window. The evaporator is positioned so that it provides cool air to the inside, and the condenser and compressor are mounted in the part of the unit that sits outside the window.

G. The motor that turns the condenser fan.

TRUE OR FALSE

Place a *T* or *F* in the blank to indicate if the statement is true or false.

Score __________

1. ____ The Y terminal on the thermostat is used to control the air-conditioning outdoor system.
2. ____ In the split-system air conditioner the condenser fan can run without the compressor running.
3. ____ The evaporator fan can run without the compressor running.
4. ____ The evaporator fan is located in the outdoor unit in a split-system air conditioner.
5. ____ The G terminal on the thermostat is used to control the evaporator fan relay.

MULTIPLE CHOICE

Circle the letter that represents the correct answer to each question.

Score __________

1. The evaporator fan for a split-system air conditioner is a:
   a. squirrel-cage fan.
   b. blade fan.
   c. shaded-pole fan.

2. If the compressor tries to start but does not make any noise, you should suspect that the:
   a. run winding is good and drawing current but the start winding is open.
   b. main fuse is blown or the compressor is not receiving voltage.
   c. start winding is good and drawing current but the run winding is open.

3. If the condenser fan in a split system is running and the compressor will not run and is humming, you should suspect that:
   a. the main fuse is blown or the main power to the outdoor section is not receiving power.
   b. the indoor temperature is at the thermostat setpoint, so the compressor does not need to run.
   c. either the start or run winding is open or the components connected to the start windings of the condenser fan are malfunctioning.

4. If the evaporator fan in a window unit is operating correctly and little or no air is moving across the condenser fins, you should suspect that the:
   a. condenser fan is faulty or not running at all.
   b. condenser fan blade may be loose or the condenser fins may be covered with dirt, since the condenser and evaporator use the same fan motor.
   c. fuse to the condenser fan is probably open.

5. If the condenser fan in a packaged-type air-conditioning system is not running and the compressor and evaporator fan are running, you should suspect that:
   a. the condenser fan has an open winding or a wire to the fan is open.
   b. one of the main fuses to the packaged unit is open.
   c. there is a short in the compressor motor, which is causing low voltage in the condenser motor.

LAB EXERCISE: OBSERVING AND UNDERSTANDING AIR-CONDITIONING SYSTEMS

Safety for this Lab Exercise

In this lab exercise you will be requested to take the covers and doors off of several different types of HVAC systems so that you can locate and identify different parts. You must be aware that the power may be turned on to the systems so that you can observe their normal operation. You should be aware of any open electrical terminals, of motors that are turning a fan or other mechanical device such as belts and pulleys if they are used, and to stay clear of all moving parts.

Tools and Materials Needed to Complete the Lab Exercise

Your instructor will provide a number of HVAC systems such as a window air conditioner, a split-system air conditioner, and a packaged air-conditioning system for you to observe their operation and identify their basic parts. Your instructor may provide some of the basic parts of these systems for you to check on a workbench. The parts on the workbench should not have any power applied so you can handle them safely and take a closer look at them.

References to the Text

Refer to Chapter 15 in the textbook for additional information. You may need to read sections of the chapter again to help you understand the material in this exercise.

Sequence to Complete the Lab Task

Identifying the Parts of a Window Air Conditioner

Be sure to check with your instructor for the steps where you are requested to locate parts. Indicate that you have found the parts and can explain what they are and how they work.

1. Figure 15–1 shows a typical window type air conditioner. Check the window air conditioner your instructor has provided and locate the fan motor, which has a double-ended motor that has the evaporator fan on one end and the condenser fan on the other end, the compressor and the evaporator coil, and condenser coil.

2. Turn power on to the window air conditioner and notice that cool air comes out from the evaporator coils and warm air comes out of the condenser fan. Check the air from the evaporator and condenser fan.

3. The electrical diagram for the window air conditioner is shown in Figure 15–2. In the diagram, you can see the window air conditioner has a combination switch that allows you to select a number of speeds for the fan. Turn the fan switch control to switch the fan speed to slow, medium, and high speeds. Did you notice the fan speed change? __________

4. The electrical diagram also shows a thermostat that allows you to set the temperature that brings the compressor on and off. Adjust the thermostat on your unit to the highest temperature possible. Did the compressor turn off? __________

5. Did the evaporator/condenser fan continue to run, or did it turn off? __________

6. Wait three minutes and turn the thermostat to its lowest temperature setting. Did the compressor turn on again? __________

7. You are called to troubleshoot a window air conditioner. The evaporator/condenser fan is running, but the compressor will not turn on. List all the things you would troubleshoot or you would suspect the problems could be.

   ____

   ____

8. Your instructor may provide an evaporator/condenser fan motor on a workbench so you can get a better look at it. Check out the fan and answer the following questions.
   a. If the evaporator fan is producing airflow across the evaporator coil, should the condenser fan be turning? __________
   b. If the evaporator fan is running on the slowest speed, at what speed will the condenser fan be running? __________
   c. Can the evaporator fan continue running when the compressor cycles off?

      ____

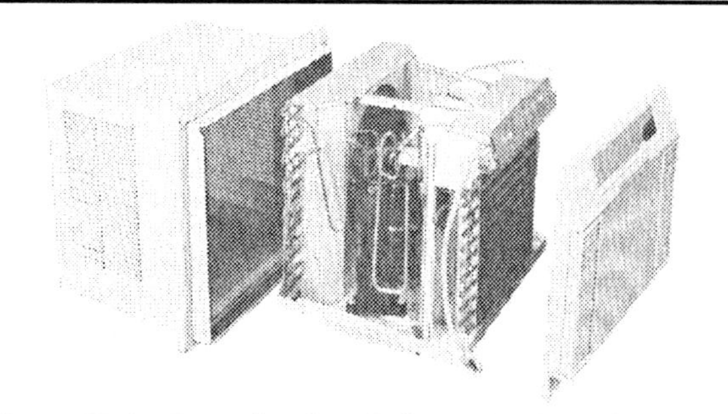

Figure 15–1 Example of a window unit air conditioner. (*Courtesy of Comfort-Aire Heat Controller Inc.*)

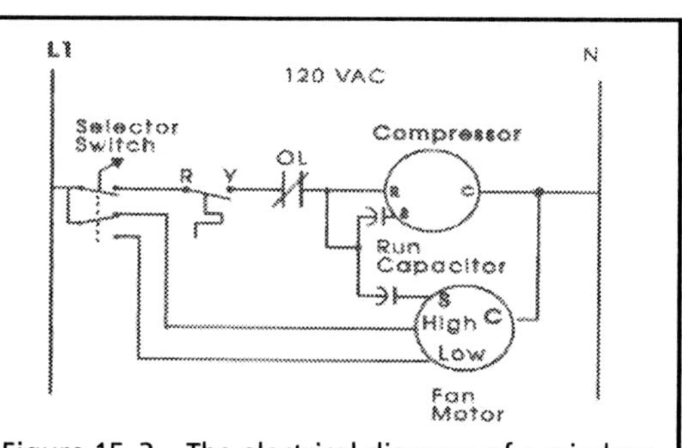

Figure 15–2 The electrical diagram of a window air conditioner.

Sequence to Complete the Lab Task

Identifying the Parts of a Split-System Type Air-Conditioning System

Be sure to check with your instructor for the steps where you are requested to locate parts. Indicate that you have found the location of the parts and can explain what they are and how they work.

The split-system air conditioner has the compressor and condenser fan in the outdoor unit, which is called the condenser, and the evaporator fan is mounted in the indoor unit. The indoor unit may be a furnace or it might be an air handler, which would only have the evaporator coil and evaporator fan. Figure 15–3 shows the parts of a split system. Figure 15–3a shows the condenser unit, Figure 15–3b shows the indoor unit, and Figure 15–3c shows a cutaway diagram of the condenser unit.

1. Check the condenser unit that your instructor has provided. Locate the following parts and explain the function of each.

   a. Compressor

   ____

   b. Condenser fan

   ____

   c. Condenser coils

   ____

   d. The electrical controls, including the compressor contactor.

   ____

2. Check the indoor unit and locate the following components and explain the function of each.

   a. Evaporator fan

   ____

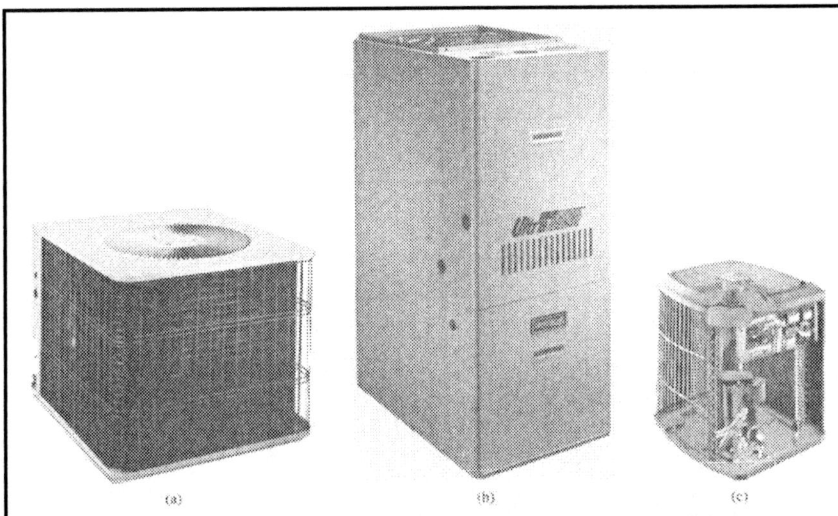

(a) (b) (c)

Figure 15–3 Condenser and indoor unit for split-system air conditioner. *(Courtesy of Armstrong Air Conditioning Inc., United Technologies Carrier Corporation)*

b. Transformer

___

c. Evaporator coils

___

3. Turn power on to the split system and operate the indoor and outdoor units. Answer the following questions.

   a. Where is the thermostat located to operate the split-system air conditioner?

   ___

   b. Where is the transformer located to provide the control voltage for the system?

   ___

   c. Can the indoor fan run when the compressor is turned off? ___

   d. Can the indoor fan run when the outdoor fan is turned off? ___

   e. If the compressor is running, can the outdoor fan be turned off? ___

Sequence to Complete the Lab Task

Identifying the Parts of a Packaged-Type Air Conditioner

Be sure to check with your instructor for the steps where you are requested to locate parts. Indicate that you have found the location of the parts and can explain what they are and how they work.

The packaged system type air conditioner has the compressor, condenser fan, and condenser coil in one side of the unit, and the evaporator fan is mounted in the other end of the packaged unit. The packaged system type air conditioner is also called a rooftop unit because one of its applications is to be mounted on the roof of a store or office where air is ducted directly through the roof into the ceiling of the room that is being cooled. Some packaged units have a heating system, which is gas fuel or electric heating coils. Figure 15–4a shows a picture of a packaged system type air conditioner, and Figure 15–4b shows a cutaway diagram of the packaged system type unit. The packaged-type air conditioner is similar to a large window unit, except it has separate motors for the evaporator fan, which moves indoor air and the condenser fan.

1. Locate the packaged air conditioner that your instructor has provided. Locate and identify the following parts, and explain the function of each.

   a. Compressor

   ___

   b. Condenser fan

   ___

   c. Condenser coils

   ___

   d. The electrical controls, including the compressor contactor

   ___

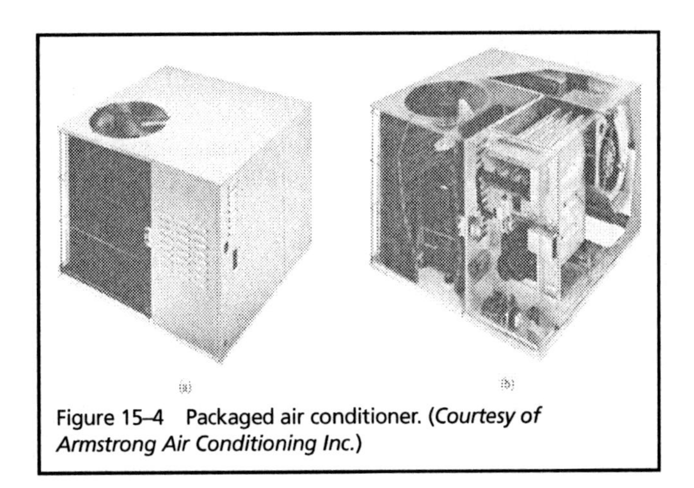

Figure 15–4 Packaged air conditioner. (*Courtesy of Armstrong Air Conditioning Inc.*)

2. Check the section of the packaged unit that has the indoor fan and locate the following components explain the function of each.
   a. Evaporator fan
   ____
   b. Transformer
   ____
   c. Evaporator coil
   ____
3. Turn power on to the packaged-type air conditioner and operate it in the cooling mode. Answer the following questions.
   a. Where is the thermostat located to operate the packaged-type air conditioner?
   ____
   b. Where is the transformer located to provide the control voltage for the system?
   ____
   c. Can the indoor fan run when the compressor is turned off? __________
   d. Can the indoor fan run when the outdoor fan is turned off? __________
   e. If the compressor is running, can the outdoor (condenser) fan be turned off?
   ____

Checking Out

When you have completed this lab exercise, clean up your area, return all tools and supplies to their proper place, and check out with your instructor. Your instructor will initial here to indicate you are ready to check out. _______

CHAPTER 16

Electrical Control of Heat Pump Systems

OBJECTIVES

At the end of this lab exercise you will be able to:

1. Identify the basic parts of a split-system type heat pump system.
2. Identify the basic parts of a packaged system type heat pump system.
3. Explain the operation of a heat pump system.
4. Identify the signals from the heat pump thermostat.
5. Explain the operation of the heat pump reversing valve.

INTRODUCTION AND OVERVIEW

This lab exercise will show you pictures of a variety of heat pump equipment. Your instructor will provide similar equipment in your lab to check out and identify the basic parts. You will identify the basic parts and understand the operation of the heat pump system so you can learn about troubleshooting them, and be able to explain their operation to their owners when you are making a troubleshooting call as an HVAC technician.

TERMS

Backup heat
Cooling cycle
Heating cycle
Heat pump
Reversing valve
Second-stage heating

MATCHING

Place the letter A–F for the definition from the list that matches with the terms that are numbered 1–6.

Score __________

1. ____ Backup heat
2. ____ Cooling cycle
3. ____ Heat pump
4. ____ Heating cycle
5. ____ Reversing valve
6. ____ Second-stage heating

A. The part of the heat pump cycle where heat is added to the indoor space.
B. The heating source that is added to a heat pump and is used to provide heat when the heat pump cannot provide enough heat, or when it goes into defrost mode.
C. The second level of heating for a furnace or heat pump. In a heat pump system, the second stage of heating is an electrical heating coil. In a gas-burning furnace, the second stage may allow additional burners or more gas to be used in the ignition process. In an electrical furnace, the second stage of heating is an additional electrical heating coil.

D. An HVAC system that has a special reversing valve that allows the system to provide cooling to the indoor space in the summer, and reverse the refrigerant flow to provide heating to the indoor space in the winter. The indoor coil acts as an evaporator in the cooling cycle during the summer, and it acts as a condenser in the heating cycle during the heating season.

E. Referring to a heat pump, this cycle is when the system is running in the air-conditioning mode and liquid refrigerant is pumped to the indoor coil, which provides cool air to the indoor space.

F. A special solenoid that is used in a heat pump that reverses the refrigeration flow from heating mode to cooling mode. This valve routes the liquid refrigerant to the indoor coil during the cooling cycle, and it routes the liquid refrigerant to the outdoor coil during the heating cycle.

TRUE OR FALSE

Place a *T* or *F* in the blank to indicate if the statement is true or false.

Score __________

1. ____ When the heat pump is in the cooling cycle, the indoor coil will be cool.
2. ____ When the heat pump is in the defrost cycle, the outdoor coil will be cool.
3. ____ When the heat pump is in the defrost cycle, the auxiliary heat will be on.
4. ____ If the reversing valve has an open in its coil and it becomes inoperative, the heat pump will remain in the heating cycle.
5. ____ When the heat pump is in the defrost cycle, the outdoor fan will be de-energized, and the indoor fan will be running.

MULTIPLE CHOICE

Circle the letter that represents the correct answer to each question.

Score __________

1. When the heat pump is in the cooling cycle, _______________ will be warm.
   a. the indoor coil
   b. the outdoor coil
   c. both coils

2. When the heat pump is in the defrost cycle, _______________ will be warm.
   a. the indoor coil
   b. the outdoor coil
   c. both coils

3. When the heat pump is in the emergency heat cycle, the:
   a. compressor will be turned off and the auxiliary heat will be turned on.
   b. outdoor section will be energized, and the indoor section will be turned off.
   c. reversing valve must be in the cooling mode.

4. The auxiliary heat will be energized:
   a. only when the heat pump is in the defrost mode.
   b. only when the heat pump is in the heating mode and second-stage heating is needed.
   c. when the heat pump is in the defrost mode or when it needs additional second-stage heating in the heating mode.

5. The balance point is the:
   a. outdoor temperature at which it is no longer economical to operate the heat pump in the heating cycle, and the auxiliary heat is used when the temperature is below this point.
   b. indoor temperature at which it is no longer economical to operate the heat pump in the heating cycle, and the auxiliary heat is used when the temperature is below this point.
   c. comparison of the indoor temperature and the outdoor temperature at which the heat pump is the most efficient.

LAB EXERCISE: OBSERVE AND UNDERSTAND HEAT PUMP

Safety for this Lab Exercise

In this lab exercise you will be requested to take the covers and doors off of several different types of heat pump systems so that you can locate and identify different parts. You must be aware that the power may be turned on to the systems so that you can observe their normal operation. You should be aware of any open electrical terminals, of motors that are turning a fan or other mechanical device such as belts and pulleys if they are used, and to stay clear of all moving parts.

Tools and Materials Needed to Complete the Lab Exercise

Your instructor will provide a number of heat pump systems such as a split-system heat pump and a packaged type heat pump system for you to observe their operation and identify their basic parts. Your instructor may provide some of the basic parts of these systems for you to check out on a workbench. The parts on the workbench should not have any power applied so you can handle them safely and take a closer look at them.

References to the Text

Refer to Chapter 16 in the textbook for additional information. You may need to read sections of the chapter again to help you understand the material in this exercise.

Sequence to Complete the Lab Task

Identifying the Parts of a Split-System Type Heat Pump System

Be sure to check with your instructor for the steps where you are requested to locate parts. Indicate that you have found the location of the parts and can explain what they are and how they work.

The biggest difference between the heat pump and the typical air-conditioning system is that the heat pump has a reversing valve that can reverse the flow of refrigerant so that in the heating cycle, the refrigerant is pumped from the compressor as a warm vapor to the indoor coil where the indoor fan transfers the heat from the indoor coil to the indoor living space. When the system is in this mode for heating, the outdoor unit becomes the evaporator coil, which pulls heat out of the air. When the heat pump is in the heating mode, the indoor coil becomes the condenser and the outdoor coil becomes the evaporator. When the heat pump is in the air-conditioning mode, the indoor coil is the evaporator and the outdoor coil is the condenser, just like a normal air-conditioning system. The reversing valve takes care of routing the refrigerant to the proper coils during each cycle.

Another part of the heat pump system that makes it different from the air-conditioning system is that the heat pump will create frost on the outdoor coil when it is acting as an evaporator during the heating season. Since the outdoor coil is operating as an evaporator coil during the heating cycle, it will be several degrees cooler than the surrounding ambient temperature. This means that when the temperature is around 40° or cooler, the outdoor coil will be cool enough to create a layer of frost on it. The frost will begin to coat the outdoor coil and slow the movement of air across the coil and begin to make it less efficient. As more frost is added, the more inefficient the coil becomes, until it reaches the point where it needs to be defrosted. When the heat pump goes into the defrost mode, the reversing valve directs the refrigerant flow so that the warm refrigerant goes to the outdoor coil just like when it is in the air-conditioning mode. This will cause the frost to melt off the coil very quickly. During the time the heat pump is in defrost mode, the indoor coil will act as an evaporator and cool the air that moves across this. Since the system is in heating mode, the cool air is not desirable, so backup heat must be energized to warm the air so it is acceptable to the people in the heated space. The defrost cycle can be terminated by the temperature on the coil, the pressure in the outdoor coil, or by a timing mechanism.

The heat pump can efficiently transfer heat from air down to about 36°F. This is called the balance point, and depending on the cost of electricity or other fuels such as natural gas, propane gas, or fuel oil, the balance point in some areas of the country will be a little warmer or a little cooler. In the colder parts of the country, a backup source of heat will be needed when the heat pump can no longer draw heat efficiently from the outdoor air during the heating season. The secondary source of heat is called *backup heat* or *second-stage heat,* and it can be electric resistance heat, natural gas, propane, or fuel oil.

The heat pump thermostat will be more complex than a normal air-conditioning thermostat, and a later part of this lab exercise will help you understand all the parts to the heat pump thermostat. The heat pump thermostat is normally mounted on the wall in the conditioned space.

1. The split-system heat pump system has the compressor and condenser fan in the outdoor unit. In the heat pump, this part of the system is called the outdoor unit rather than the condenser, since the indoor unit will act as the condenser during the heating cycle. The fan that is mounted in the indoor unit is called the indoor fan rather than the evaporator fan, since the outdoor unit acts as the evaporator during the heating cycle. The indoor unit may be a furnace or it might be an air handler, which would only have the coil and fan. Figure 16–1 shows examples of split-system heat pumps. Figure 16–1a shows the indoor unit, and Figure 16–1b shows the outdoor unit. Check the outdoor unit that your instructor has provided. Locate the following parts and explain the function of each.

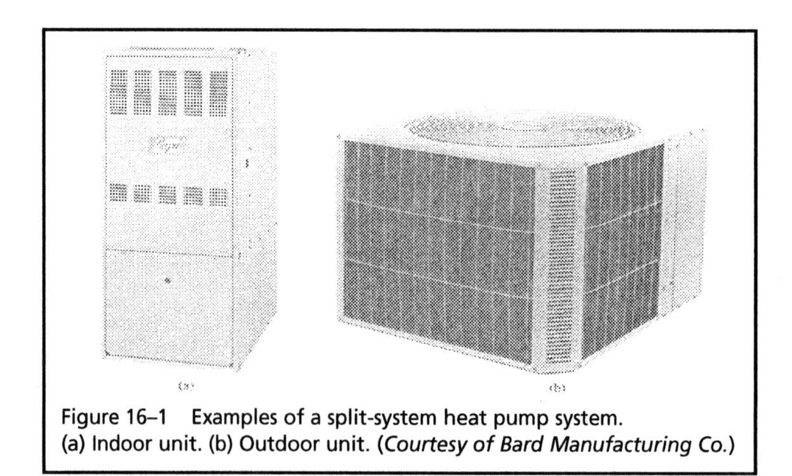

Figure 16–1 Examples of a split-system heat pump system.
(a) Indoor unit. (b) Outdoor unit. (*Courtesy of Bard Manufacturing Co.*)

a. Compressor

____

b. Outdoor fan

____

c. Outdoor coils

____

d. Reversing valve

____

e. The electrical controls including the compressor contactor

____

2. Check the indoor unit and locate the following components, and explain the function of each.

a. Indoor fan

____

b. Transformer

____

c. Indoor coil

____

d. Second-stage heating system (electric coils, gas furnace, or oil furnace)

____

3. Turn power on to the split-system heat pump in air-conditioning mode. Make sure the indoor unit and outdoor unit are both operating. Answer the following questions.

a. Is the air coming from the indoor unit warm or cool? __________

b. Is the air coming across the outdoor unit warm or cool? __________

c. Is the temperature of the air moving across the indoor and outdoor units similar to a normal air-conditioning system? __________

d. Where is the thermostat located to operate the split-system heat pump? __________

e. Where is the transformer located to provide the control voltage for the system?

____

f. Can the indoor fan run when the compressor is turned off? __________

g. Can the indoor fan run when the outdoor fan is turned off? __________

h. If the compressor is running, can the outdoor fan be turned off? __________

Sequence to Complete the Lab Task

Identifying the Reversing Valve and Explaining Its Operation

The heat pump has a reversing valve that switches the direction of the refrigerant from the heating cycle to the cooling cycle. Your instructor will locate a reversing valve for you to work with and observe its parts up close. Figure 16–2 shows a cutaway picture of a reversing valve. Figure 16–3a shows the reversing valve connected in the cooling cycle, and Figure 16–3b shows the reversing valve in the heating cycle. You can see when the reversing valve is in the cooling cycle, the hot gas from the compressor is routed to the outdoor coil, and the evaporator coil is connected to the suction line of the compressor. When the reversing valve is in the heating mode, the hot gas from the compressor is routed to the indoor coil and the outdoor coil is connected to the suction line of the compressor.

1. Locate the reversing valve in a heat pump and trace the lines that come from the compressor high-pressure line. Next, locate the line that goes to the outdoor coil, and then find the line that goes to the indoor coil. Finally, locate the line that goes to the compressor suction line. After you have identified all of the lines, call your instructor and discuss this information.

2. If the heat pump is in the air-conditioning system, explain how you can tell if the reversing valve is in the correct position and is moving refrigerant to the correct coil.

   ____

   ____

3. If the heat pump is in the heating mode, explain how you can tell if the reversing valve is in the correct position and is moving refrigerant to the correct coil.

   ____

   ____

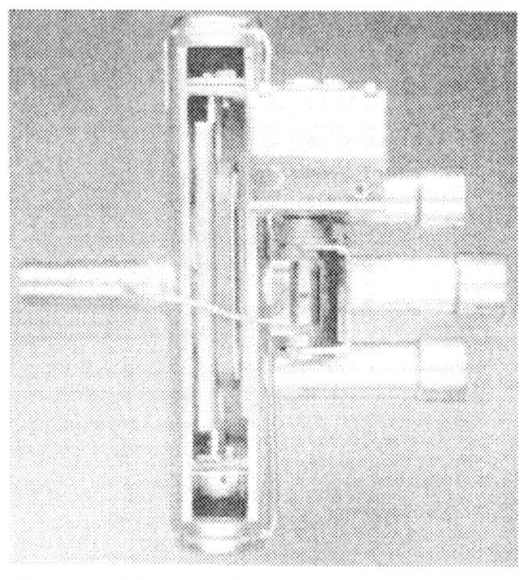

Figure 16–2 A reversing valve is a specialized valve that is used in heat pumps. The inlet on the left side is from the compressor discharge. The middle of the three connections on the right side is connected to the compressor suction line. The outside two connections go to the indoor coil and the outdoor coil. *(Courtesy of Alco Controls Division, Emerson Electric Company)*

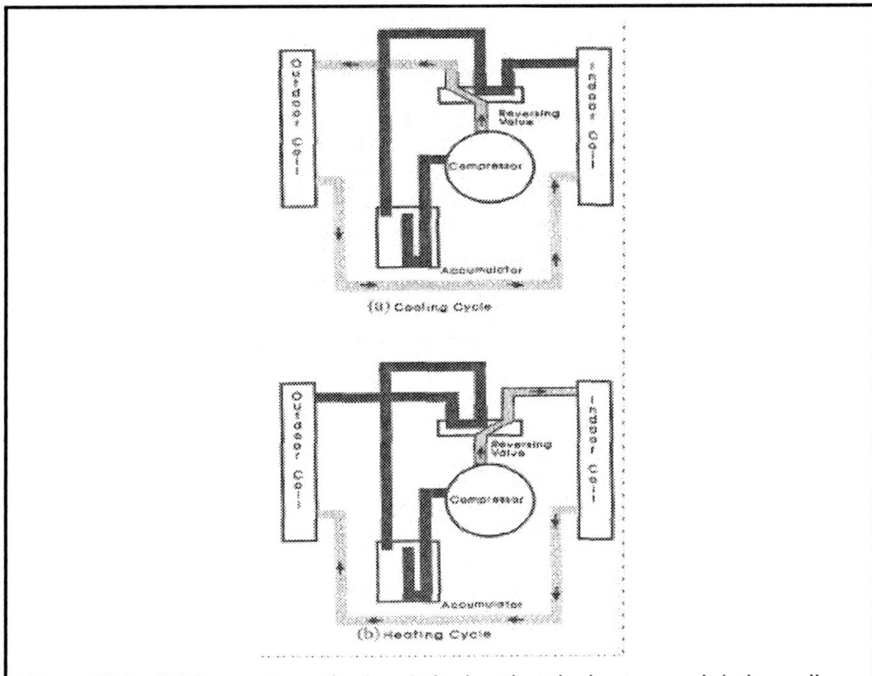

Figure 16–3 (a) A reversing valve is switched so that the heat pump is in its cooling cycle. The indoor space is cooled when the heat pump is in this mode. (b) The reversing valve is switched so that the heat pump is heating the indoor space.

4. Since the reversing valve is actually a solenoid coil, explain how you could test the coil to see if it has continuity.

____

____

5. Since the reversing valve is a solenoid valve, could you hold a metal screwdriver close to the coil when it is energized to determine if it is being attracted by the magnetic field in the coil? __________

Sequence to Complete the Lab Task

Identifying the Terminals and Operation of the Heat Pump Thermostat

The thermostat that controls the heat pump is a little more complex than a typical heat/cool thermostat because it must provide additional signals to the reversing valve. The system also needs a secondary source of heat that is used as a backup heat, also used during the defrost cycle. Figure 16–4 shows a heat pump connected to a heat pump thermostat. Figure 16–5 shows a list of the signals the heat pump thermostat has and all the devices that are connected to each terminal. Your instructor will provide a heat pump thermostat with its subbase that is not connected to any system, so you can look at it close up and identify all the terminals. You will also be provided a heat pump thermostat that is connected to an operating heat pump so you can observe the system's operation.

When you learn about the heat pump thermostat you must think about the way the system operates when it is in cooling mode (air conditioning), heating mode, second-stage heating, and defrost mode. When the system is in air-conditioning mode, the reversing valve is not energized

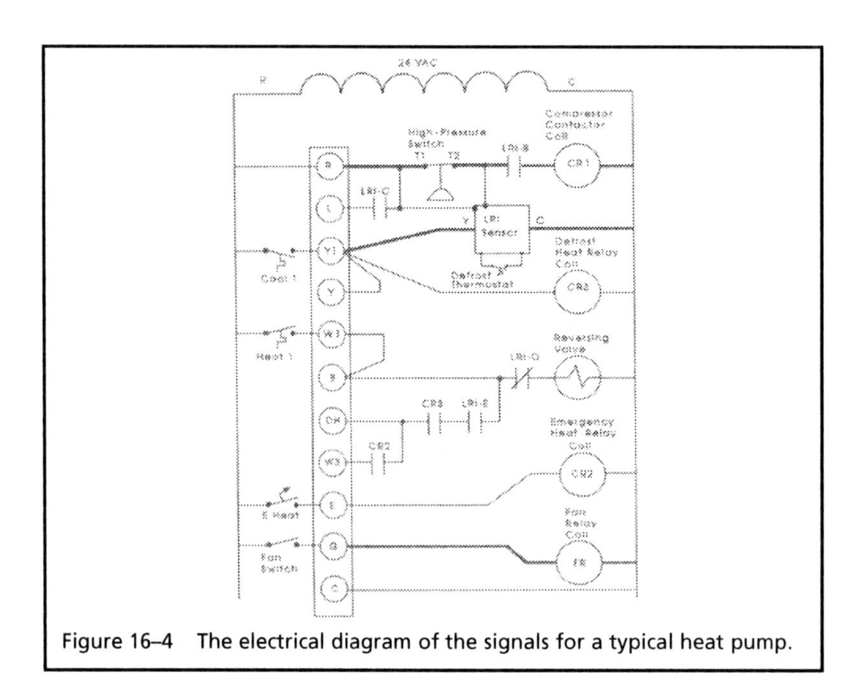

Figure 16–4 The electrical diagram of the signals for a typical heat pump.

and it directs the warm gas from the compressor to the outdoor unit, which operates as the condenser. The liquid refrigerant is directed to the indoor coil that operates as the evaporator. During this cycle the thermostat sends a signal to the compressor contactor from the Y1 terminal. The compressor and outdoor fan receive voltage when the contacts of the compressor contactor close. The indoor fan will receive a signal through terminal G when the thermostat calls for cooling, or if the fan switch is in the on position.

During the heating cycle, a heating circuit in the thermostat sends a signal out terminal W1, which goes to the coil of the reversing valve and the compressor contactor. When the reversing valve receives a signal, it changes position so that it directs the warm gas from the compressor to the indoor coil, which now operates as a condenser coil and condenses the refrigerant to a liquid and gives up heat. The liquid from the indoor coil is directed to the outdoor coil, which operates as an evaporator, which transfers heat from the air around the coil into the refrigerant. If the indoor space does not become warm enough with only the heat pump running and the indoor temperature drops 2° below the heating setpoint temperature, the second stage of the thermostat will become energized and call for stage 2 heating, which sends a signal to W2. The second-stage heating can be an electric heating coil, a gas furnace, or oil furnace. The heat pump continues to operate when the second-stage heating is energized, unless the outdoor temperature is too cold to allow the system to be efficient. When the indoor space warms up to within 2° of the heating setpoint, the second stage of heat drops out and the system continues to run on just the heat pump.

During the defrost cycle, the system is switched to operate as though it is in air-conditioning mode. This sends the warm gas from the compressor to the outdoor coil, and the warm temperature of the refrigerant should be warm enough to melt the frost from the coils. Since the indoor coils act like an evaporator when the system is in the air-conditioning mode, cool air will be sent into the conditioned space, so the backup heat source will need to be energized during the defrost cycle to warm the indoor air up. The defrost cycle can be terminated by time, temperature, or any other means that can determine when the frost is removed.

Figure 16–5 shows a table of all the signals for the heat pump. Your instructor will provide a heat pump thermostat for you to work with. Write the letters for each signal in the table and identify what the signal does. Have your instructor check your work.

TERMINAL DESIGNATION DESCRIPTIONS

Terminal Designation	Description
L	Heat pump malfunction indicator for systems with malfunction connection
O	Changeover valve for heat pump energized constantly in cooling
B	Changeover valve for heat pump energized constantly in heating
Y	Compressor Relay
Y2	2nd Stage Compressor
W/E	Heat Relay/Emergency Heat Relay (Stage 1)
W2	2nd Stage Heat (3rd Stage Heat in HP2)
G	Fan Relay
RH	Power for Heating
RC	Power for Cooling
C	Common wire from secondary side of cooling system transformer or heat only system transformer

Figure 16–5 Terminal designation descriptions for heat pump systems. (*Courtesy of White- Rodgers*)

Signal Terminal Designation	Description

Sequence to Complete the Lab Task

Identifying the Parts of a Packaged Type Heat Pump

Be sure to check with your instructor for the steps where you are requested to locate parts. Indicate that you have found the location of the parts and can explain what they are and how they work.

The packaged system type heat pump has the compressor, outdoor fan, and outdoor coil in one side of the unit, and the indoor fan is mounted in the other end of the packaged unit near the indoor coil. The packaged system type heat pump is also called a rooftop unit because one of its applications is to be mounted on the roof of a store or office where air is ducted directly through the roof into the ceiling of the room that is being cooled. The packaged heat pump system has a backup heating system, which is electric heating coils, natural gas, propane, or fuel oil. Figure 16–6 shows a picture of a packaged system type heat pump, and you can see this unit has electric heating coils as the backup heat.

1. Locate the packaged heat pump that your instructor has provided, and locate and identify the following parts and explain the function of each.

   a. Compressor

   ____

   b. Outdoor fan

   ____

   c. Outdoor coils

   ____

   d. Reversing valve

   ____

   e. The electrical controls, including the compressor contactor

   ____

2. Check the indoor unit and locate the following components and explain the function of each.

   a. Indoor fan

   ____

   b. Transformer

   ____

   c. Indoor coil

   ____

   d. Second-stage heating system (electric coils, gas furnace, or oil furnace)

   ____

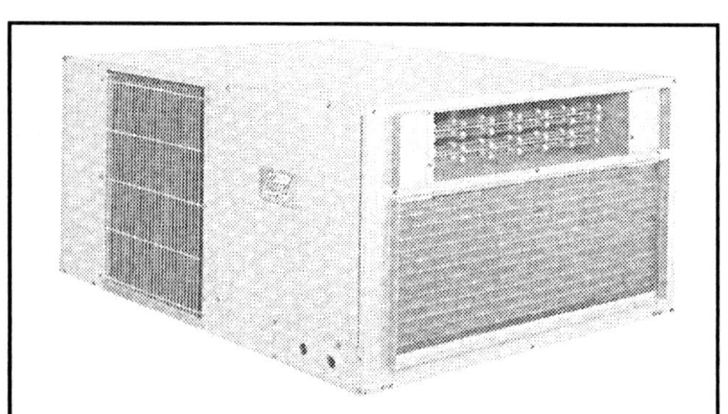

Figure 16–6 A packaged-type heat pump. (*Courtesy of Bard Manufacturing Co.*)

3. Turn power on to the packaged type heat pump in air-conditioning mode. Make sure the indoor unit and outdoor unit are both operating. Answer the following questions.
   a. Is the air coming from the indoor unit warm or cool? __________
   b. Is the air coming across the outdoor unit warm or cool? __________
   c. Is the temperature of the air moving across the indoor and outdoor units similar to a normal air-conditioning system? __________
   d. Where is the thermostat located to operate the packaged type heat pump? __________
   e. Where is the transformer located to provide the control voltage for the system?
   ____
   f. Can the indoor fan run when the compressor is turned off? __________
   g. Can the indoor fan run when the outdoor fan is turned off? __________
   h. If the compressor is running, can the outdoor fan be turned off? __________

Sequence to Complete the Lab Task

Identifying the Parts of a Ground Source or Geothermal Heat Pump

Your instructor may be able to take you to see an installation of a ground source heat pump. The ground source heat pump may be called a geothermal heat pump. If you cannot visit a ground source heat pump you may be able to find pictures of a system in your area or on the Internet. The ground source heat pump is generally used where the winter temperatures are colder and the heating season is longer. This type of heat pump does not have an outdoor unit; rather, it has a heat exchanger and a large amount of tubing that has antifreeze solution flowing through it. The tubing is buried in the ground or in the bottom of a pond or lake where it can transfer heat from the surrounding area. In the winter the antifreeze solution is circulated through the tubing that is buried in the ground and it collects heat from its surrounding area. Over the course of the heating season this cools the dirt or water that surrounds the tubing, which makes the system more efficient in the first months of the cooling season. The tubing is buried below the frost line so that it can transfer the most amount of heat energy during the winter. A typical heat pump is only efficient down to about 35° to 40°F because the amount of heat energy in the air is less, and at that point the system frosts over frequently and must be defrosted, which is the most inefficient time of the cycle. The ground source heat pump takes advantage of the fact that the earth below the frost line stays at approximately 50° all year long, and this provides enough heat energy to make the heat pump operate efficiently. The 50° earth also ensures that the system will never need to be defrosted.

During the cooling (air conditioning) season, the flow of the fluid moves heat from the system to the ground, which again is about 50°. This means that the fluid that flows through the earth will be about 50° and it will also warm the ground. The earth or water in a pond is a very efficient storage medium that can store large amounts of energy. At the end of the cooling season in September, the earth surrounding the tubing has been warmed all summer and is slightly warmer than the earth would normally be. This causes the ground source heat pump to be more efficient when it changes over to its heating cycle in the fall, and since fluid cools the earth during the winter, the earth is slightly cooler than normal when the system switches over to air-conditioning cycle in the spring.

The ground source heat pump is generally used for heating the home, and also helping with providing hot water for the home. In the summer the temperature in the coil is warm enough to provide hot water for the home without any additional heat source, and in the winter the water is warmed to the point that only a small amount of additional heating source is needed. This combination makes the heating, air-conditioning, and hot water system very efficient. The initial installation cost and the cost of the ground source heat pump is more expensive than an air-to-air heat pump, but the payback is quick and will last a long period of time, which makes the system a very efficient heating and cooling system.

Checking Out

When you have completed this lab exercise, clean up your area, return all tools and supplies to their proper place, and check out with your instructor. Your instructor will initial here to indicate you are ready to check out. _______

CHAPTER 17

Electrical Control of Refrigeration Systems

OBJECTIVES

At the end of this lab exercise you will be able to:

1. Identify the basic parts of a refrigeration system.
2. Identify the basic parts of an ice maker.
3. Explain the operation of a refrigeration system.
4. Explain the operation of an ice maker.

INTRODUCTION AND OVERVIEW

This lab exercise will show you pictures of a variety of refrigeration equipment. Your instructor will provide similar equipment in your lab to check out. You will identify the basic parts and understand the operation of refrigeration system so you can learn about troubleshooting them and be able to explain their operation to their owners when you are making a troubleshooting call as an HVAC technician. The major difference between HVAC equipment and refrigeration equipment is that the evaporator coil on some systems are allowed to get below freezing, so a defrost activity is required periodically so that the coil can be cleared of frost. This lab exercise will provide activities to allow you to check out refrigeration equipment.

TERMS

Condenser
Defrost control
Door heater
Electronic defrost control
Electronic refrigeration control
Evaporator
Freezer display case
Ice maker
Oil pressure switch
Pressure control
Reach-in cooler
Temperature control
Temperature defrost
Timed defrost
Walk-in cooler

MATCHING

Place the letter A–L for the definition from the list that matches with the terms that are numbered 1–12.

Score ________

1. ____ Condenser
2. ____ Defrost control
3. ____ Door heater
4. ____ Electronic defrost control
5. ____ Electronic refrigeration control
6. ____ Evaporator
7. ____ Freezer display case
8. ____ Ice maker
9. ____ Pressure control
10. ____ Reach-in cooler
11. ____ Temperature control
12. ____ Timed defrost

A. A control for defrost that has an electronic circuit that provides the timer function rather than an electromechanical timer motor.
B. The outdoor unit for an air-conditioning system or refrigeration system. The compressor, condenser coil, and compressor fan are mounted in this unit. The condenser coil allows the refrigerant vapor to cool down to a point that it turns back to a liquid form.
C. A sensor that is attached to a switch that can detect the increase or decrease in refrigerant or air pressure. The pressure control can also be used to control the temperature of a refrigeration system since refrigerant pressure can be used to determine temperature at the coil.
D. The cooling coil in an air-conditioning or refrigeration system. Liquid refrigerant flows into this coil where it changes from a liquid to a vapor as it absorbs heat.
E. A specialized case that allows refrigerated products to be displayed in a way that consumers can view and reach into it to select products while the products stay at the refrigerated temperature.
F. An electrical component that controls one or more sets of contacts by time, temperature, or pressure to ensure a defrost cycle occurs. This control is used in refrigeration and heat pump systems to ensure frost buildup on the evaporator coil is kept to a minimum.
G. A control for refrigeration functions that provides the time functions through an electronic type timer instead of an electromechanical timer motor.
H. A system that has a specific job of creating cube or flake ice for residential or commercial consumption.
I. A special heating strip that is built into the door gasket that is mounted on the door of a refrigerator or freezer.
J. A case used for frozen products in grocery stores or other stores. The case can maintain temperatures below 32°F.
K. Defrost cycle for a refrigeration system or heat pump that is brought on by a timer control. The defrost cycle also turns off with a timer.
L. A control that turns an air-conditioning or refrigeration system on at temperature setpoints.

TRUE OR FALSE

Place a *T* or *F* in the blank to indicate if the statement is true or false.

Score __________

1. ____ A cooler or freezer can use a low-pressure control as a thermostat to control the temperature in the display case.
2. ____ The door heaters on a reach-in cooler are needed because the door handles are too cold when people touch them.
3. ____ In most reach-in coolers, the evaporator fan and condenser fans are three-phase motors.
4. ____ Freezers require a defrost heater because frost will build up on evaporator fins and block air flow.
5. ____ The ice maker shown in the electrical diagram in Figure 17–11 uses a hot-gas solenoid to warm the evaporator plate during the harvest cycle.

MULTIPLE CHOICE

Circle the letter that represents the correct answer to each question.

Score __________

1. If you are troubleshooting a reach-in cooler because its compressor is not running, you should test the:
   a. door heaters because an open in a heater will keep voltage from reaching the compressor.
   b. supply voltage, compressor relay, and thermostat to see that voltage is passing through these components and reaching the compressor.
   c. door switches to ensure that voltage is reaching the compressor.

2. The compressor in the electrical diagram for the reach-in cooler in Figure 17–2 is connected as a:
   a. capacitor-start, induction-run (CSIR) compressor.
   b. permanent split-capacitor (PSC) compressor.
   c. split-phase compressor with a current relay.

3. The condenser fan motor in the electrical diagram for the reach-in cooler in Figure 17–2 is connected as a:
   a. shaded-pole motor.
   b. capacitor-start, induction-run (CSIR) motor.
   c. permanent split-capacitor (PSC).

4. The compressor and condenser fan for the reach-in cooler in Figure 17–1 is physically located:
   a. outside, since this is a split-system type unit.
   b. below the cooler case toward the back of the unit.
   c. on top of the cooler case.

5. If the defrost cycle for the freezer in Figure 17–9 will not activate, you should suspect that:
   a. the defrost timer motor has stopped.
   b. one of the door switches is faulty.
   c. the drain line heater has failed, and the drain is frozen shut.

LAB EXERCISE: OBSERVING AND UNDERSTANDING REFRIGERATION SYSTEMS

Safety for this Lab Exercise

In this lab exercise you will be requested to take the covers and doors off of several different types of refrigeration systems so that you can locate and identify different parts. You must be aware that the power may be turned on to the systems so that you can observe their normal operation. You should be aware of any open electrical terminals, of motors that are turning a fan or other mechanical device such as belts and pulleys if they are used, and to stay clear of all moving parts.

Tools and Materials Needed to Complete the Lab Exercise

Your instructor will provide a number of refrigeration systems such as a display case or reach-in cooler for you to observe their operation and identify their basic parts. Your instructor may provide some of the basic parts of these systems for you to check out on a workbench. The parts on the workbench should not have any power applied so you can handle them safely and take a closer look at them.

References to the Text

Refer to Chapter 17 in the textbook for additional information. You may need to read sections of the chapter again to help you understand the material in this exercise.

Sequence to Complete the Lab Task

Identifying the Parts of a Display Case Type Refrigeration System

Be sure to check with your instructor for the steps where you are requested to locate parts. Indicate that you have found the location of the parts and can explain what they are and how they work.

The biggest difference between the typical air-conditioning system and a refrigeration system is that the refrigeration system is designed to operate at lower temperatures, which may be cold enough to create frost on the evaporator, which must be defrosted from time to time. Figure 17–1 shows a typical refrigerated display case. The temperature for the display case may be below freezing if it is used to store frozen food, and the temperature may be 35° to 40°F if the system is used to store refrigerated products such as milk, cheese, or meat products.

1. Your instructor will provide you with a refrigeration system to check out. Locate the compressor, condenser coil, condenser fan, and the evaporator. If the system is a walk-in cooler, the system will have an evaporator fan to move air across the evaporator coils. If the system is a freezer unit, it will only have an evaporator plate, and there may not be an evaporator fan. The system may have a thermostat to control the temperature, or a low-pressure switch may be used. If the system reaches temperatures where frost is created, it will have a defrost method built in. The defrost process can be accomplished by directing hot gas through the evaporator during the defrost cycle, or it may be accomplished by turning on electric defrost heaters to provide enough heat to cause the frost to melt. The refrigeration system must also have heaters built into the door seals and other places around the system where the buildup of ice or frost may cause problems with moisture or ice and frost being created. Locate each component on the system and identify the function of each.

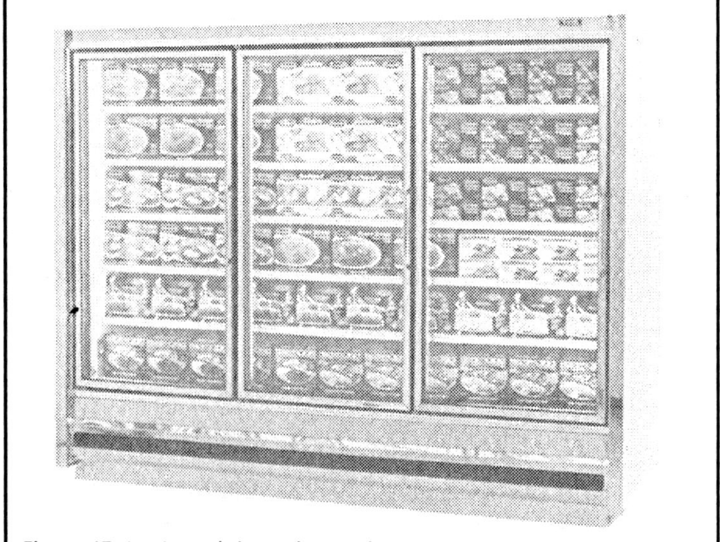

Figure 17–1 A reach-in cooler used in grocery stores or convenience stores. (*Courtesy of Tyler Refrigeration Corporation*)

a. Compressor

____

b. Condenser coils

____

c. Condenser fan

____

d. Temperature control

____

e. Evaporator coils

____

f. Evaporator fan, if present

____

g. Defrost mechanism

____

h. Door heaters or other heaters

____

2. Turn power on to the refrigeration system and allow it to run for 10 to 20 minutes. Make sure the unit has been operating long enough that the system is starting to cool down. Answer the following questions.

   a. Is the evaporator coil warm or cool? __________

   b. Is the air coming across the outdoor unit warm or cool? __________

   c. Is the temperature of the air moving across the evaporator and condenser similar to the operation of an air-conditioning system? __________

   d. What type of temperature control does your system have? __________

   e. Does your system have any door heaters or other types of heaters? Where are they located? __________

Sequence to Complete the Lab Task

Identifying the Electrical Diagram for a Refrigeration System

1. Figure 17–2 shows an electrical diagram of a refrigeration system. List all the loads in this system diagram.

   a. ______________________________

   b. ______________________________

   c. ______________________________

d. ______

e. ______

f. ______

g. ______

2. What is the device that controls the temperature for this refrigeration system? ______

3. How many thermostats does this system have? ______

4. How many heaters are in this system? ______

5. What are the names of the heaters?

______

6. What device controls the defrost cycle? ______

7. How many door switches does this unit have? ______

8. What gets turned off if any of the doors are opened? ______

9. Is the compressor protected against high or low pressure? ______

10. How many capacitors does the compressor use to get started? ______

11. When does the condenser fan run?

______

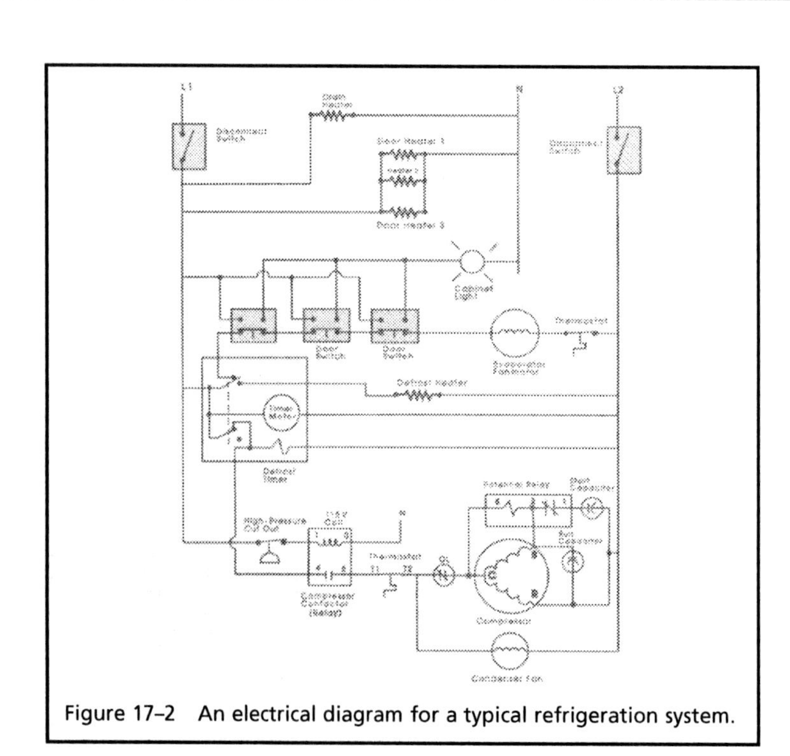

Figure 17–2 An electrical diagram for a typical refrigeration system.

Sequence to Complete the Lab Task

Identifying the Parts of an Ice Maker

Be sure to check with your instructor for the steps where you are requested to locate parts. Indicate that you have found the location of the parts and can explain what they are and how they work.

An ice maker is a specifically designed refrigeration system that takes clean water and creates ice cubes that are fit for human consumption. Figure 17–3 shows examples of an ice maker that has a specially designed evaporator plate that creates ice cubes in a specific shape for the machine. The ice maker has a harvest cycle that is used to ensure the ice cubes come free from the evaporator and drop into the storage bin, which is also refrigerated at a temperature below freezing to keep the ice cubes fresh and ready for service. When you are working on an ice maker, you can troubleshoot it in much the same way as you would an HVAC system in that it has a compressor, a condenser coil, a condenser fan, an evaporator plate, a thermostat, a water solenoid, and possibly a water pump for constant water flow. Your instructor will provide you with an ice maker to check out. Locate the compressor, the condenser coil, condenser fan, and the evaporator. This system should have a source of water and a storage bin for the ice, which you will also need to locate.

1. Figure 17–4 shows an electrical diagram of an ice maker. List all the loads in this ice maker diagram.
   a. ____________________
   b. ____________________
   c. ____________________
   d. ____________________
   e. ____________________
   f. ____________________
   g. ____________________
2. What is the device that controls the temperature for this refrigeration system? __________
3. How many thermostats does this system have? __________

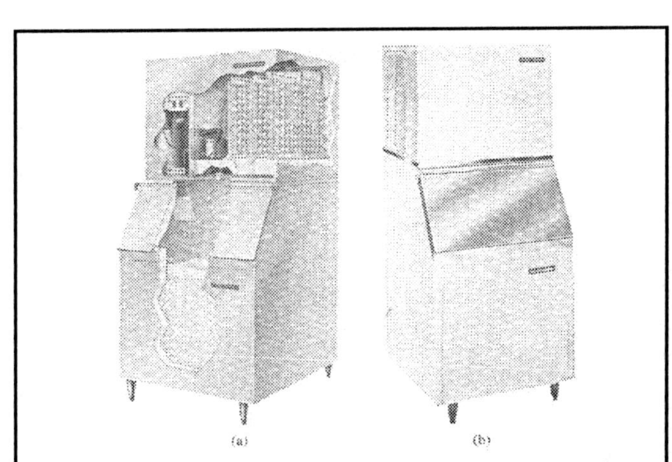

Figure 17–3 (a) A cutaway picture of a cube-type ice machine. (b) A cube-type ice machine. (*Courtesy of Scotsman*)

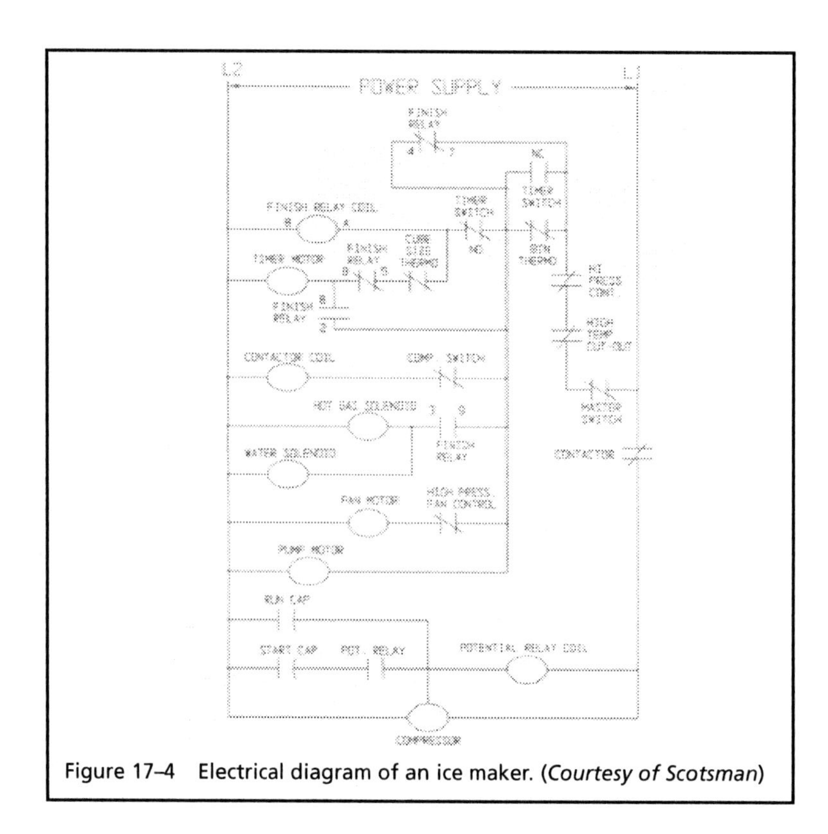

Figure 17–4 Electrical diagram of an ice maker. (*Courtesy of Scotsman*)

4. How many solenoids are in this system? __________

5. What are the names of the solenoids?

   ____

6. What device controls the defrost cycle? __________

7. What controls the on/off operation of the compressor? __________

8. Is the compressor protected against high or low pressure? __________

9. How many capacitors does the compressor use to get started? __________

10. When does the condenser fan run?

    ____

Checking Out

When you have completed this lab exercise, clean up your area, return all tools and supplies to their proper place, and check out with your instructor. Your instructor will initial here to indicate you are ready to check out. _______

CHAPTER 18

Troubleshooting Heating, Air-Conditioning, and Refrigeration Systems

OBJECTIVES

At the end of this lab exercise you will be able to:

1. Understand the process of troubleshooting an HVAC or refrigeration system.
2. Use the voltage drop method of troubleshooting to locate an open in an electrical circuit.
3. Troubleshoot an air-conditioning system.
4. Troubleshoot an electric furnace.

INTRODUCTION AND OVERVIEW

This lab exercise will show you an electrical diagram of an HVAC system and an electric furnace. You will learn to use the voltage drop method of troubleshooting. Troubleshooting is a process of identifying which circuits in a system are working correctly, and then which circuit is causing the problem. After you identify the circuit that is causing the problem, then you can do a voltage loss procedure to identify exactly where in the circuit you have the problem and where voltage is lost.

The main thing to remember about troubleshooting is that you need to use a procedure that can be used every time that you troubleshoot. This lab exercise will provide a chance for you to learn about the troubleshooting procedure and exactly how to execute the steps of the procedure to locate the problem in the circuit every time you use it.

TERMS

Circuit breaker
Common point on motor
Compressor contactor
Continuity test
Megger
Ohmmeter
Run winding
Start capacitor
Start winding
Troubleshooting
Voltage drop method of troubleshooting
Voltmeter

MATCHING

Place the letter A–L for the definition from the list that matches with the terms that are numbered 1–12.

Score __________

1. ____ Anti short cycle timer (ASCT)
2. ____ Circuit breaker
3. ____ Compressor contactor
4. ____ Continuity test
5. ____ Megger
6. ____ Ohmmeter
7. ____ Run winding
8. ____ Start capacitor
9. ____ Start winding
10. ____ Troubleshooting
11. ____ Voltage drop method of troubleshooting
12. ____ Voltmeter

A. A megohm meter that produces high voltage and low current. The main purpose of this meter is to measure the resistance of insulation in coils of motors and transformers to determine the breakdown of resistance, which may indicate problems due to the deterioration of insulation or the presence of moisture in windings.
B. An electrical device specifically designed to protect against overcurrent and can be reset.
C. A process of finding fault in an electrical or HVAC system. The troubleshooting process works best when you are able to identify everything in the system that is working correctly, and then only checking the remaining part of the circuit that is not working correctly.
D. An electrical meter that is designed to measure voltage. This meter can be analog (with a needle) or digital (with a numerical readout display).
E. A timer that is specifically designed to keep a compressor from cycling on and off too quickly. After a compressor turns off, the timer will prevent it from coming back on too quickly, and allows the system pressure to drop to a level that will not cause the motor to draw excessive current.
F. A relay that has contacts that can carry current in excess of 15 amps. This device has a coil that controls a number of normally open contacts that are wired in series with the compressor motor windings.
G. A troubleshooting procedure that provides the means for locating an open circuit in an electrical circuit by applying voltage to the circuit and making a number of voltage measurements. One terminal of the meter must be placed on the return line of the power supply and left there for the duration of the tests. The test procedure then specifies that voltage is checked at each terminal point in the circuit in sequence starting at the power supply. The point where voltage is lost indicates the point where the open circuit exists.
H. A meter that is designed to measure resistance and indicate the value in ohms.
I. One of two windings in a split-phase motor that only stays in the circuit for a few seconds. This winding in a split-phase motor has a higher resistance than the run winding.
J. A test that measures the resistance of components such as switch contacts, fuses, or wires.
K. A capacitor that is connected in the start winding of a compressor or other single-phase motor. This capacitor is mounted in a plastic case and is only in the circuit for a few seconds.
L. The main winding of a single-phase motor or compressor that has less resistance than the start winding.

TRUE OR FALSE

Place a *T* or *F* in the blank to indicate if the statement is true or false.

Score __________

1. ____ When you test a component with an ohmmeter, you should always ensure the power is turned off.
2. ____ When you test a component with a voltmeter, you should always ensure the power is turned off.
3. ____ When you are called to troubleshoot an HVAC or refrigeration system, you should use all of your senses to determine what parts of the HVAC or refrigeration system are operating correctly.
4. ____ You could use your "sense of feel" to determine if a fan or other part of the system is vibrating excessively.
5. ____ A clamp-on ammeter could be used to determine which of the three lines that supply voltage to a motor has a blown fuse if the motor is humming and trying to start.

MULTIPLE CHOICE

Circle the letter that represents the correct answer to each question.

Score __________

1. A wiring diagram shows:
   a. the sequence of operation for the circuit.
   b. the location of each component in the circuit.
   c. how each component operates.
   d. how an HVAC or refrigeration system should be troubleshooted.

2. A ladder diagram shows:
   a. the sequence of operation for the circuit.
   b. the location of each component in the circuit.
   c. how each component operates.
   d. how an HVAC or refrigeration system should be troubleshooted.

3. Troubleshooting may best be described as:
   a. finding the part of the HVAC or refrigeration system that is broken.
   b. identifying the parts of an HVAC or refrigeration system that are working correctly, and through a process of elimination determining what is faulty and not working correctly.
   c. swapping parts until the HVAC or refrigeration system starts running again.
   d. removing parts from an HVAC or refrigeration system and testing each one with an ohmmeter.

4. The control circuit includes:
   a. the motor and other loads in a circuit.
   b. the motor and other controls in a circuit.
   c. switches and other controls in a circuit.
   d. All of the above

5. The load circuit includes:
   a. the motor and other loads in a circuit.
   b. the motor and other controls in a circuit.
   c. switches and other controls in a circuit.
   d. All of the above

LAB EXERCISE: UNDERSTANDING TROUBLESHOOTING PROCEDURES

Safety for this Lab Exercise

In this lab exercise you will be requested to take the covers and doors off of several different types of air conditioners and furnaces. You will be provided several diagrams that you will work through and use for your troubleshooting procedure. Then your instructor will provide an HVAC system with an open circuit in it so that it will not run. You will be requested to use the voltage loss procedure to locate the problem. You must be aware that the power will be turned on to the systems so that you can make voltage measurements to locate the open in the circuit. You should be aware of any open electrical terminals, of motors that are turning a fan or other mechanical device such as belts and pulleys if they are used, and to stay clear of all moving parts.

Tools and Materials Needed to Complete the Lab Exercise

Your instructor will provide an HVAC system that has an open in one of its circuits so that the system will not run. Your instructor will need to work closely with you since you will have power turned on to the system to use the voltage loss method to find the problem. Your instructor will also provide an electric furnace with a problem in its circuit so that it will not work. Again, you must have supervision from your instructor to complete the troubleshooting process since power will be turned on while you try to locate the problem using the voltage loss troubleshooting method.

References to the Text

Refer to Chapter 18 in the textbook for additional information. You may need to read sections of the chapter again to help you understand the material in this exercise.

Sequence to Complete the Lab Task

Using an Electrical Diagram to Create a Troubleshooting Procedure

Be sure to check with your instructor when you are ready to observe the system and make your first procedure for troubleshooting. You will also have to ensure that it is safe to make the checks, and your instructor will initial the space when you are ready to make the tests where voltage will be present in the machine. Have your instructor initial here to indicate you are being supervised during this lab procedure. _______

1. Figure 18–1 shows the electrical diagram for a typical air-conditioning system. If the diagram for the HVAC system your instructor is providing is different, then you need to work from that diagram. We will use the diagram in Figure 18–1 to create the troubleshooting procedure to use the voltage loss method of troubleshooting.

   In the diagram, you can see two circuits: one controls the compressor contactor coil, and the other controls the coil of the fan relay. We will begin with a problem in the system where the compressor will not start.

   Since the troubleshooting call is that the compressor will not start, you should begin by trying to operate the system and get the compressor to run. Do this by turning on power to the system and setting the thermostat to air conditioning and set it to the lowest temperature to be sure the thermostat is "calling" for cooling. This will allow you to verify the customer's complaint.

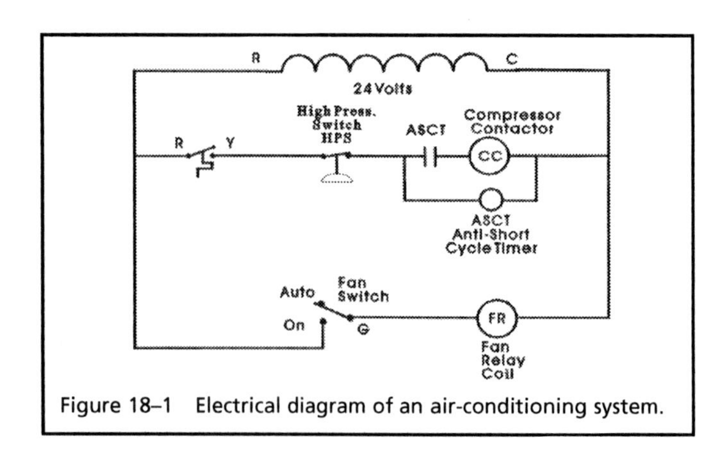

Figure 18–1 Electrical diagram of an air-conditioning system.

a. Did the compressor start? __________

b. If the compressor will not start, which circuit will you start troubleshooting first?

____

c. Which circuit can you ignore at this point? __________

d. Explain what has to occur in the circuit that provides voltage to the compressor contactor coil for the coil to receive its voltage.

____

____

e. Explain why you must know what provides voltage to the coil of the compressor contactor to be able to troubleshoot the circuit.

____

____

2. Figure 18–2 shows the same diagram as in Figure 18–1, but shows each test point in the circuit numbered 1–11. We will refer to these numbers during the troubleshooting process. Start the troubleshooting process by taking a voltage measurement from test points 1–2. This will indicate the voltage present at terminals R and C of the transformer.

   a. How much voltage do you measure at terminal points 1–2? __________

   b. If you measure 24 volts at points 1–2 (terminals R and C), what does it indicate?

   ____

   ____

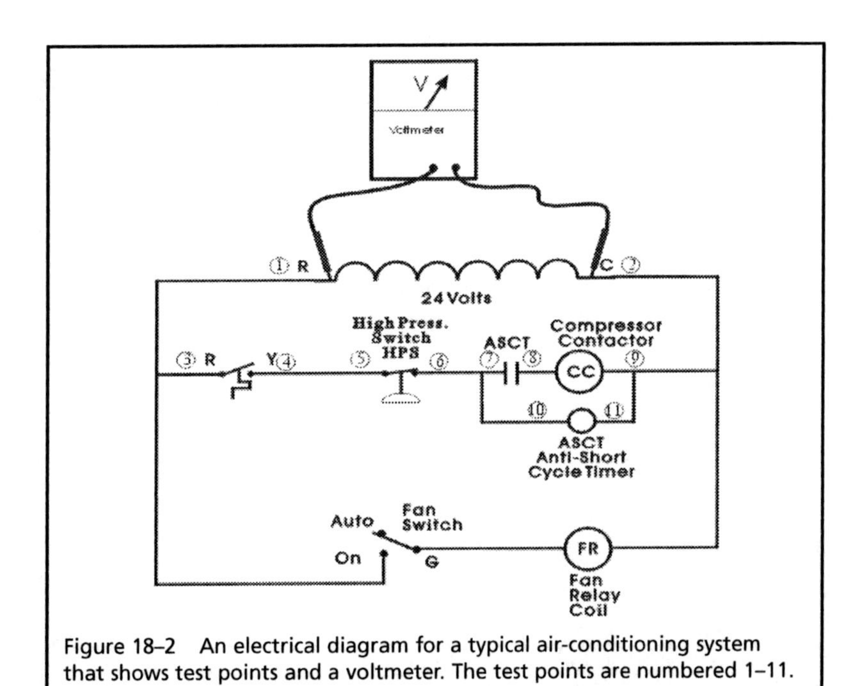

Figure 18–2 An electrical diagram for a typical air-conditioning system that shows test points and a voltmeter. The test points are numbered 1–11.

c. If you measure zero volts at points 1–2 (terminals R and C), what does it indicate?

____

____

d. If you have 24 volts at points 1–2, you can move to the next step in the process. If you have zero volts, you have found the problem (the transformer is not producing voltage at the secondary).

3. Leave one of your meter probes at terminal 2 and move the other probe to test point 3.

   a. How much voltage do you measure at terminal points 3–2? __________

   b. If you measure 24 volts at points 3–2 (terminals R at the thermostat to C at transformer), what does it indicate?

   ____

   ____

   (*NOTE:* You may not be able to make this test if the thermostat is some distance from your unit such as at a residence. If this is the case, you would skip to the next test point and only come back to this test point if you measure zero volts.)

   c. If you measure zero volts at points 3–2 (terminals R at the thermostat and C at the transformer), what does it indicate?

   ____

   ____

   d. If you have zero volts, you have found the problem (the wire between the transformer terminal R and the thermostat terminal R has an open and must be replaced). If you have 24 volts at points 3–2, you can move to the next step in the process.

4. Leave one of your meter probes at terminal 2 and move the other probe to test point 4. Test point 4 is terminal Y that you can locate on the terminal board in the indoor unit. Terminal Y is where the wire comes from the thermostat back to the terminal block of the indoor unit.

   a. How much voltage do you measure at terminal points 4–2? __________

   b. If you measure 24 volts at points 4–2 (terminals Y at the indoor unit to C at transformer), what does it indicate?

   ____

   ____

   c. If you measure zero volts at points 4–2 (terminal Y at the indoor unit and terminal C at the transformer), what does it indicate?

   ____

   ____

   d. If you have zero volts, you have found the problem (the wire between the transformer terminal R and the thermostat terminal R has an open and must be replaced). If you have 24 volts at points 4–2, you can move to the next step in the process.

5. Leave one of your meter probes at terminal 2 and move the other probe to test point 5. Test point 5 is the terminal connected to the left side of the high-pressure switch.
   a. How much voltage do you measure at terminal points 5–2? ________
   b. If you measure 24 volts at points 5–2 (terminal at the left side of the high-pressure switch to C at transformer), what does it indicate?

      ____

      ____

   c. If you have zero volts at points 5–2, you have found the problem (the wire between terminal Y at the indoor unit and the terminal on the left side of the high-pressure switch has an open and must be replaced). If you have 24 volts at points 4–2, you can move to the next step in the process.

6. Leave one of your meter probes at terminal 2 and move the other probe to test point 6. Test point 6 is the terminal on the right side of the high-pressure switch.
   a. How much voltage do you measure at terminal points 6–2? ________
   b. If you measure 24 volts at points 6–2 (terminal at the right side of the high-pressure switch to C at transformer), what does it indicate?

      ____

      ____

   c. If you have zero volts at points 6–2, you have found the problem (the high-pressure switch is open. It might be open because of a high-pressure condition, or it might be a problem with the switch and it must be replaced). If you have 24 volts at points 6–2, you can move to the next step in the process.
   d. What conditions should you check that could cause the high-pressure switch to trip?

      ____

      ____

7. Leave one of your meter probes at terminal 2 and move the other probe to test point 7. Test point 7 is the terminal on the left side of the anti short cycle timer.
   a. How much voltage do you measure at terminal points 7–2? ________
   b. If you measure 24 volts at points 7–2 (terminal at the left side of the anti short cycle timer to C at transformer), what does it indicate?

      ____

      ____

   c. If you have zero volts at points 7–2, you have found the problem (the wire between terminal on the right side of the high-pressure switch and the terminal on the left side of the anti short cycle timer has an open in it and must be replaced). If you have 24 volts at points 7–2, you can move to the next step in the process.

8. Leave one of your meter probes at terminal 2 and move the other probe to test point 8. Test point 8 is the terminal on the right side of the contacts of the anti short cycle timer.
   a. How much voltage do you measure at terminal points 8–2? __________
   b If you measure 24 volts at points 8–2 (terminal on the right side of the contacts of the anti short cycle timer to C at transformer), what does it indicate?

   ____

   ____

   c. If you have zero volts at points 8–2, you have found the problem (the contacts in the anti short cycle timer is open). This might be a condition where the timer is in its time delay cycle, which is usually about three to five minutes. If you have waited longer than five minutes and the contacts are still open, you should look for a reset button. If there isn't one, you can turn off power to the HVAC unit and then turn it back on, which will also reset the timer. If the contacts remain open, you will have to replace the anti short cycle timer. If you have zero volts at points 8–2, you can move to the next step in the process.

9. If you have voltage at terminal 8 to C, then you have determined that the entire circuit on the left side of the compressor contactor coil is okay. If the compressor contactor coil is not energized and does not pull its contacts in when you have voltage all the way to terminal 8, you should move the focus of your test to the part of the circuit on the right side of coil. For this part of the test, you should leave one voltmeter probe on terminal 8, and move the terminal that was on terminal C to test point 9, which is the terminal on the right side of the compressor contactor coil. Now you have the meter probes directly across the coil of the compressor contactor.
   a. How much voltage do you measure at test points 8–9? __________
   b. If you measure zero volts, then the problem is the wire between test point 2 at terminal C of the transformer, and the terminal on the right side of the compressor contactor coil is bad. You will need to replace the wire to provide power to the coil.
   c. What does it mean if you measure 24 volts at terminals at test points 8 and 9 directly across the coil terminals of the compressor contacts?

   ____

   ____

   d. If you measure 24 volts across the coil terminals but the coil does not pull in the contacts, you will need to turn off power to the HVAC unit, disconnect the wires from the compressor contactor coil, and test the coil for continuity. What is the resistance that you measure in the coil of the contactor? __________

10. If the coil is bad, you will need to replace the contactor. This completes the troubleshooting tests for this circuit. Explain why you need to use a procedure like this every time you troubleshoot an electrical circuit that will not energize a coil to find where you have lost voltage.

    ____

    ____

Sequence to Complete the Lab Task

Troubleshooting an Electric Furnace

Figure 18–3 shows an electrical diagram of an electric furnace. In this exercise you will be requested to determine the test points that you would use to determine if different parts of the circuit are not energizing.

1. The first heating bank is not energizing. Identify the points you would test to determine why the first heating bank is not energizing.

   a. What causes the first bank of heat to energize?

   ____

   ____

   b. List the test points to determine if you have voltage in this circuit.

   ____

   ____

2. The first bank of heat energizes but the second bank will not energize.

   a. What causes the second bank of heat to energize?

   ____

   ____

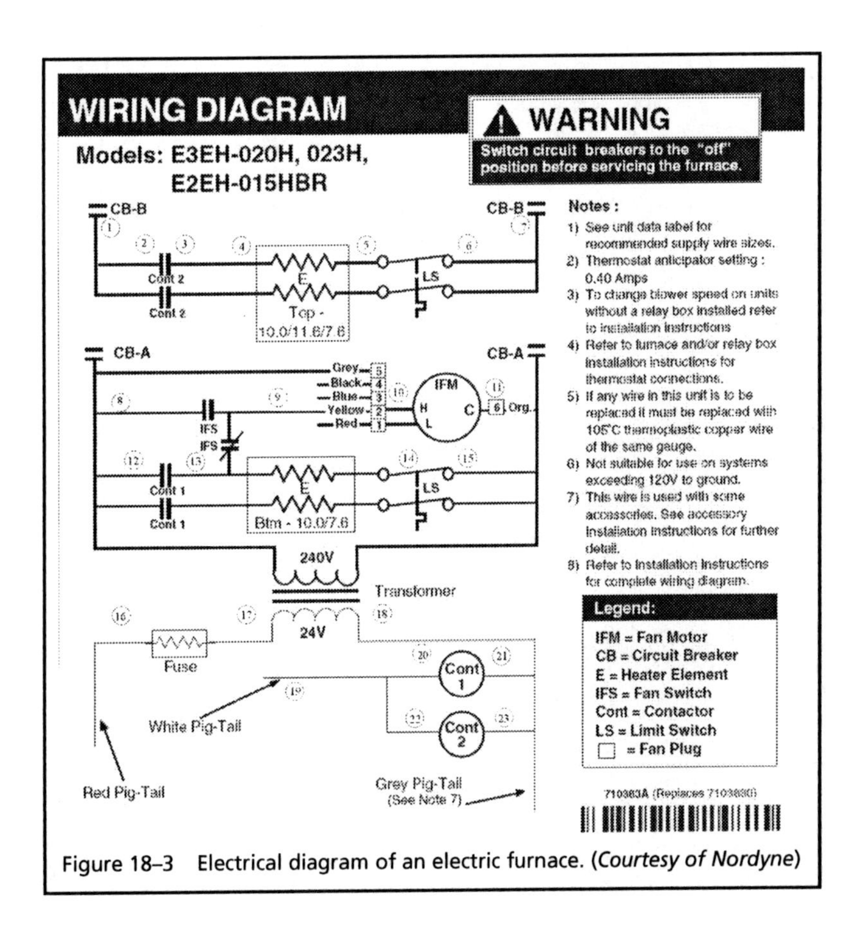

Figure 18–3 Electrical diagram of an electric furnace. (*Courtesy of Nordyne*)

b. List the test points to determine if you have voltage in this circuit.

____

____

3. The internal fan motor (IFM) does not turn on.

   a. What causes the fan motor to become energized?

   ____

   ____

   b. List the test points to determine if you have voltage to the fan motor.

   ____

   ____

   c. If you have voltage at the high or low speed terminal of the motor and the C terminal, what does it indicate?

   ____

   ____

   d. If you suspected the fan motor is not functioning, explain how you would test its windings.

   ____

   ____

   e. Do you need to turn off the power to make the test you suggest in the previous step?

   ____

4. Your instructor may provide an electric furnace for you to troubleshoot. A problem will be put into the electric furnace so that it will not run. You will need your instructor's supervision if you troubleshoot a furnace with power applied. Have your instructor initial here to indicate you have supervision. _______

Checking Out

When you have completed this lab exercise, clean up your area, return all tools and supplies to their proper place, and check out with your instructor. Your instructor will initial here to indicate you are ready to check out. _______

CHAPTER 19

Electronic Devices for HVAC Systems

OBJECTIVES

At the end of this lab exercise you will be able to:

1. Understand the opertion of half-wave and full-wave rectifiers in the power supply for electronic circuits used in HVAC systems.
2. Understand the opertion of variable-frequency drives used to control the variable speed fans and pumps.
3. Troubleshoot the power supply of an electronic circuit board.

INTRODUCTION AND OVERVIEW

Some HVAC technicians feel they do not need to know anything about electronics since they will not be repairing electronic circuit boards. You will find that it is very important to understand the function of electronic circuits in air conditioners, furnaces, and refrigeration systems so that you can troubleshoot and replace electronic boards with confidence. This lab exercise will show you basic electronic devices commonly found in air conditioners, furnaces, and refrigeration systems. As these systems have become more efficient, they have integrated more electronic circuits and devices into their operations. As a technician, you are not expected to troubleshoot electronic components; rather, you will need to understand how electronic circuits function and do the same job as a basic electrical mechanic so that you can troubleshoot a system and know when you should remove and replace boards. If you do not understand the function of the electronic boards, you will not feel comfortable about changing them and you may change out boards that are not bad. You will also need to know basic theory about electronic circuits, or you will not feel comfortable about working on the newest electronic circuits. The basic knowledge of electronics will also help you if you go for additional schooling that the equipment manufacturers provide after you are on the job. You will also find that newer, higher efficiency air conditioners and furnaces now use variable-frequency drives (VFD), which allow fan motors, compressors, and pump motors to run at variable speeds that are less than 100%, which allow these systems to be more efficient. You will learn in this chapter how the VFD changes the amount of frequency delivered to the AC motor, which will change its speed.

TERMS

Anode
Atom
Base
Cathode
Collector
Conductor
Diac
Diode
Electrons
Emitter
Full-wave bridge rectifier
Half-wave bridge rectifier
Insulator
Lattice structure
Light-emitting diode (LED)
NPN material
N-type material
Operational amplifier
PN junction
PNP material
Proportional control
P-type material
Rectifier
Semiconductor
Silicon
Silicon controlled rectifier (SCR)
Solid-state relay (SSR)
Transistor
Triac
Unijunction transistor (UJT)
Valence electron
Variable-frequency drive (VFD)

MATCHING

Place the letter A–EE for the definition from the list that matches with the terms that are numbered 1–31.

Score __________

1. ____ Anode
2. ____ Atom
3. ____ Base
4. ____ Cathode
5. ____ Collector
6. ____ Conductor
7. ____ Diac
8. ____ Diode
9. ____ Electron
10. ____ Emitter
11. ____ Full-wave bridge rectifier
12. ____ Half-wave rectifier
13. ____ Insulator
14. ____ Lattice structure
15. ____ Light-emitting diode (LED)
16. ____ NPN material
17. ____ N-type material
18. ____ PN junction
19. ____ PNP material
20. ____ Proportional control
21. ____ P-type material
22. ____ Rectifier
23. ____ Semiconductor
24. ____ Silicon material
25. ____ Silicon controlled rectifier (SCR)
26. ____ Solid-state relay (SSR)
27. ____ Transistor
28. ____ Triac
29. ____ Unijunction transistor (UJT)
30. ____ Valence electron
31. ____ Variable-frequency drive (VFD)

A. An electronic structure inside an atom where electrons are shared, thereby creating a very strong structure.
B. A wire that is usually made of copper or aluminum that carries electrical current.
C. The positive terminal of a diode or other electronic device.
D. Semiconductor material that has a majority of carriers, which are electrons and have a negative charge.
E. An electronic system that connects to a single-phase or three-phase AC motor and controls its speed by varying its frequency.
F. A solid-state unidirectional latching switch that has three terminals: anode, cathode, and gate. This device can convert AC current to DC current and vary the amount that flows through its anode-cathode circuit by controlling the firing angle of the gate.
G. The center terminal of an NPN or PNP transistor or other electronic device.
H. A two-terminal, solid-state semiconductor that allows current flow in one direction.
I. A unit of matter. The smallest unit of an element that consists of a nucleus that has a positive-charged proton and neutral-charged neutron.
J. Material with four valence electrons that is used extensively in electronic components such as diodes and transistors.
K. An electron or electrons in the outermost shell or orbit of an electron.
L. A three-terminal electronic device that has an emitter, collector, and base.
M. A device (usually a diode) that conducts current in only one direction, thereby transforming alternating current to direct current.
N. The negative part of an atom. Electrons are located in orbits (shells) and move around the nucleus (center) of the atom.
O. Material used to make a transistor emitter, base, and collector, where the emitter and collector are made from N-type material and the base is made of P-type material.
P. The negative terminal of a diode or other electronic device.
Q. A two-lead, solid-state PN junction device that produces a small amount of light when forward biased.
R. A two-terminal electronic device that is specifically designed to fire in both the positive and negative directions when its applied voltage reaches a predetermined amount.
S. One of the three terminals of a transistor. The collector attracts the electrons that are given off by the emitter.
T. A relay that does not have any mechanical parts or contacts. This device has a light-emitting diode (LED) that acts like the coil of a relay. The LED sends a light signal to the base of a phototransistor, which goes into conduction and passes current through its emitter and collector, much like the relay contacts.
U. A special transistor that has higher impedance than a PNP or NPN transistor. This device has three terminals: base 1, base 2, and emitter.
V. A material used in electronic components. This material has four valence electrons, which makes it electrically neutral.
W. A solid-state bidirectional-latching switch that has three terminals: main terminal 1, main terminal 2, and gate.
X. One segment of P-type material that is connected to one segment of N-type material to make a single junction. This is also the basic part of a diode.
Y. An automatic control circuit that uses only gain to calculate the output for the system.
Z. One of three terminals in a transistor. The emitter provides a source of electrons for the device.
AA. A diode rectifier that converts only one-half of an AC sine wave into DC voltage.

BB. Material used to make a transistor emitter, base, and collector where the emitter and collector are made from P-type material and the base is made of N-type material.
CC. Semiconductor material that has a majority of carriers, which are protons and have a positive charge.
DD. A rectifier that changes both the positive and the negative half-cycle of an AC sine wave to pulsing.
EE. A material that does not conduct electrical current easily. These materials have very high resistance. In electronics, an insulator material has five, six, or seven electrons in its valence shell.

TRUE OR FALSE

Place a *T* or *F* in the blank to indicate if the statement is true or false.

Score __________

1. ____ The triac provides switching similar to that of an SCR, except the triac operates in an AC circuit.
2. ____ The light-emitting diode (LED) is similar to a DC light bulb in that it has a very tiny filament.
3. ____ The main function of diodes and SCRs is to convert AC voltage to DC voltage.
4. ____ The variable-frequency drive has the ability to change the frequency of voltage sent to a motor to control the motor's speed.
5. ____ The unijunction transistor (UJT) and diac are important solid-state devices that provide pulse signals to other components to use as a firing signal.

MULTIPLE CHOICE

Circle the letter that represents the correct answer to each question.

Score __________

1. A PN junction is forward biased when:
   a. positive voltage is applied to the N-material and negative voltage is applied to the P-material.
   b. negative voltage is applied to the N-material and positive voltage is applied to the P-material.
   c. its junction has high resistance.

2. The op amp is an important solid-state device because it can:
   a. amplify very small signals from thermocouples and RTDs to larger voltages and currents.
   b. provide a negative firing pulse to the gate of an SCR.
   c. be used as a seven-segment display for numbers.

3. A circuit that has one SCR in it can control:
   a. AC voltage.
   b. DC voltage.
   c. both AC and DC voltage.

4. The transistor operates like a relay in that its:
   a. emitter is like the coil, and its base and collector are like a set of contacts.
   b. collector is like the coil, and its base and emitter are like a set of contacts.
   c. base is like a coil, and its emitter and collector are like a set of contacts.

5. The solid-state relay uses:
   a. an LED to receive a signal and a transistor to switch current.
   b. a capacitor to receive a signal and an IC to switch current.
   c. a transistor to receive a signal and an IC to switch current.

LAB EXERCISE: OBSERVING AND UNDERSTANDING ELECTRONIC POWER SOURCES AND VARIABLE-FREQUENCY DRIVES

Safety for this Lab Exercise

In this lab exercise you will be requested to take the covers and doors off of several different types of air conditioners and furnaces to check voltages or make changes to variable-frequency drives. You must be aware that the power will be turned on to these systems so that you can make voltage measurement or to observe the operation of these circuits. You should be aware of any open electrical terminals, of motors that are turning a fan or other mechanical device such as belts and pulleys if they are used, and to stay clear of all moving parts.

Tools and Materials Needed to Complete the Lab Exercise

Your instructor will provide an HVAC system that has one or more electronic circuit boards and a furnace that has a variable-frequency drive. Your instructor will need to work closely with you since you will have power turned on to the system when you make voltage measurements. You will also operate the variable-frequency drive on a system and observe the change in motor speeds as the electronic controls change the frequency to the motor.

References to the Text

Refer to Chapter 19 in the textbook for additional information. You may need to read sections of the chapter again to help you understand the material in this exercise.

Sequence to Complete the Lab Task

Measuring AC and DC Voltage on an Electronic Circuit Board

Electronic circuit boards operate on DC voltage. The main voltage to the furnace or air conditioner is AC voltage, so an electronic circuit needs to have a source of DC voltage. The half-wave rectifier consists of a single diode and it changes AC voltage to DC voltage. As a technician, you will need to verify that the board has DC voltage source. In this exercise you will learn how a diode half-wave rectifier changes AC voltage to DC voltage. Be sure to check with your instructor at the steps where you are observing the system and making voltage checks. You will also have to ensure that it is safe to make the checks, and have your instructor initial the space when you are ready to make the tests where voltage will be present in the machine. Have your instructor initial here to indicate you are being supervised during this lab procedure. _______

Every electronic circuit needs to have a source of DC voltage in order to operate. The main voltage supplied to the system is AC voltage. A diode rectifier circuit changes AC voltage to DC voltage. If the diode rectifier becomes inoperative, it will not produce DC voltage and the electronic circuit will not function. This exercise will show how a rectifier works and how you can make some simple voltage tests to determine if the rectifier is operating correctly. Figure 19–1 shows the electrical diagram of a half-wave rectifier. When AC voltage is supplied to the rectifier, the diode blocks half the AC wave and the result is a pulsing DC output voltage.

1. Your instructor will provide an electronic circuit board that is operational in a working furnace or air conditioner. Your task is to measure the AC voltage applied to the board, and then locate the terminals of the rectifier and measure the DC voltage.

   a. Set your meter to measure AC voltage. What is the amount of AC input voltage supplied to the board? __________

   b. Your instructor will indicate where the terminals on the circuit board for DC voltage are. Switch your meter to DC voltage and measure the amount of DC voltage.

   ____

2. If the board has the proper amount of AC voltage and the proper amount of DC voltage, what can you say about the rectifier circuit on the board?

   ____

   ____

3. If the board has the proper amount of AC voltage but the DC voltage is zero VDC, what can you say about the rectifier circuit on the board?

   ____

   ____

4. If the board does not have DC voltage, you will not repair it; rather, you would replace the complete board. To change the board you would need to turn off power, remove the wires to the board, and remove and replace the board. When you have replaced the board, you can turn power on again and test the board for proper operation.

5. If the electronic circuit needs more current, a four-diode, full-wave bridge rectifier can provide more current. Figure 19–2 shows the diagram of a four-diode, full-wave rectifier. If one of the four diodes fails, the rectifier will still produce voltage, but it will be approximately half the rated voltage for the board. If this occurs, the lower voltage will cause the circuit to not operate correctly and the board will need to be replaced.

   Your instructor will provide a four-diode, full-wave bridge rectifier. Check the specifications for this diode and record the AC and DC voltage ratings for the rectifier.

   a. AC voltage rating ______________

   b. DC voltage rating ______________

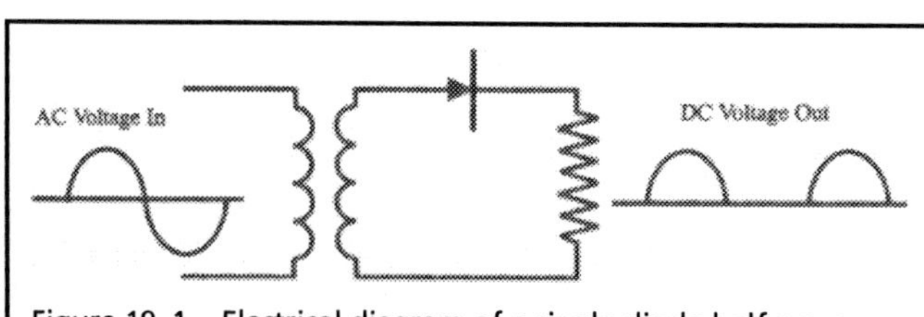

Figure 19–1 Electrical diagram of a single diode half-wave rectifier. The diode rectifier converts AC voltage to DC voltage.

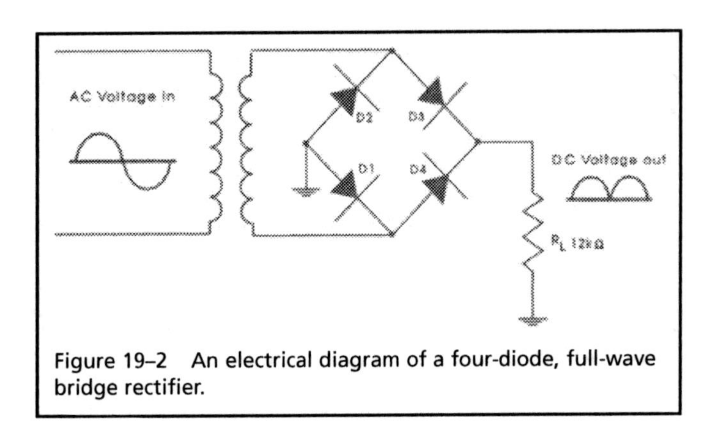

Figure 19–2 An electrical diagram of a four-diode, full-wave bridge rectifier.

6. Measure the AC voltage for this circuit. ____________

7. Measure the DC voltage for this circuit. ____________

8. If the DC voltage is less than the rated value, what can you predict is wrong with it?

____

____

9. If the DC voltage is at zero volts, what can you predict is wrong with it?

____

____

10. If the DC voltage is less than rated or at zero volts, your instructor will help you change the board. After you have replaced the board, you can try the system again.

Sequence to Complete the Lab Task

Observing the Operation of a Variable-Frequency Drive

Figure 19–3 shows an electrical diagram of a variable-frequency drive (VFD). The VFD allows the motor speed to be reduced below its rated speed. An AC motor can have its speed changed by changing the number of poles in the motor or by changing the frequency of the voltage supplied to the motor. The frequency of voltage in the United States is 60 hertz. If the voltage supplied to the motor is 60 hertz, the motor will run at its rated speed. If you reduce the frequency of the voltage, the speed will be reduced. The VFD takes 60 hertz voltage and converts the AC voltage to DC voltage by sending it through a rectifier. A bank of capacitors and inductors are used to ensure the pulsing DC is changed to straight line pure DC. The pure DC voltage is sent to a set of transistors that switch on and off to create an AC waveform. The frequency of this waveform is adjusted by the VFD to meet the demands of the HVAC system. If theVFD is controlling a fan motor, a sensor tells the VFD what speed to set the motor to. If the VFD is controlling a compressor, the amount of cooling that the system needs will cause the VFD to change the speed of the motor as it pumps refrigerant. The VFD allows the motor to run at exactly the speed that is needed, which makes the system very efficient.

1. In this exercise your instructor will provide a VFD that is controlling a motor. Set the VFD to manual mode and set it to full speed and observe the motor.

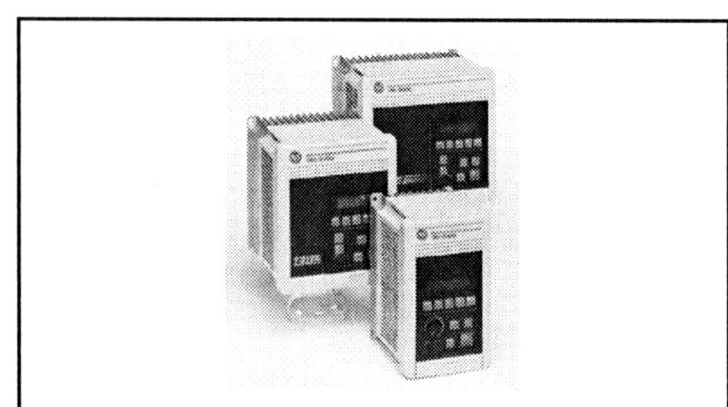

Figure 19–3 Examples of varible-frequency drives.
(*Courtesy of Rockwell Automation's Allen-Bradley Business*)

a. Is it running at full speed? __________

b. What speed does the rpm indicator on the VFD show? __________

2. Set the motor to medium speed.

   a. Did the motor slow down? __________

   b. What speed does the rpm indicator on the VFD show? __________

3. Now set the motor to a very slow speed.

   a. Did the motor slow down further? __________

   b. What speed does the rpm indicator on the VFD show? __________

4. Explain how you can tell if the VFD is not operating correctly.

   ____

   ____

Checking Out

When you have completed this lab exercise, clean up your area, return all tools and supplies to their proper place, and check out with your instructor. Your instructor will initial here to indicate you are ready to check out. _______

CHAPTER 20

Direct Digital Control (DDC) Systems and Programmable Logic Controllers (PLCs)

OBJECTIVES

At the end of this lab exercise you will be able to:

1. Understand the basic parts of the direct digital controller (DDC) system.
2. Understand the operation of DDC system that controls a water valve.
3. Understand the basic parts of the programmable logic controller (PLC).
4. Understand the opertion of PLC system that controls a solenoid valve.

INTRODUCTION AND OVERVIEW

In larger HVAC systems, computers are used to control more complex applications. When a computer is used for control it is called a direct digital controller (DDC). Newer systems use an electronic control system called a programmable logic controller (PLC) to turn on and off digital devices such as solenoids and motor starter coils, as well as analog devices such as a water valve that can open various percentages between 0 and 100%.

TERMS

A/D converter
Analog signals
ASCII
Binary-coded decimal (BCD) numbering system
Binary numbering system
Closed loop
Control loop
Controller
D/A converter
Damper motor actuator
Decimal
Digital
Direct digital controller (DDC)
Discrete
Error
Floating-point number
Hexadecimal numbering system
I/P (current over pressure) converter
Input module
Integer
Limit switch
Microprocessor
Networking
Octal numbering system
Op amp
Open loop
Output module
P/I (pressure overcurrent) converter
Pneumatic sensor
Process variable
Programmable logic controller (PLC)
Proportional, integral, and derivative (PID)
Proximity switch
Push-button switch
Scan cycle
Selector switch
Setpoint (SP)
Solid-state integrated circuit
Summing junction
Temperature indicator
Thermocouple
Transmitter
Valve actuator

MATCHING

Place the letter A–GG for the definition from the list that matches with the terms that are numbered 1–33.

Score __________

1. ____ A/D converter (analog-to-digital converter)
2. ____ ASCII
3. ____ Binary-coded decimal (BCD) numbering system
4. ____ Binary numbering system
5. ____ D/A (digital-to-analog) converter
6. ____ Damper motor actuator
7. ____ Decimal number system
8. ____ Direct digital controller (DDC)
9. ____ Error
10. ____ Floating-point number
11. ____ Hexadecimal numbering system
12. ____ I/P (current over pressure) converter
13. ____ Input module
14. ____ Integer
15. ____ Limit switch
16. ____ Microprocessor
17. ____ Networking
18. ____ Octal numbering system
19. ____ Op amp (operational amplifier)
20. ____ Open loop
21. ____ Output module
22. ____ P/I (pressure overcurrent) converter
23. ____ Process variable
24. ____ Programmable logic controller (PLC)
25. ____ Proximity switch
26. ____ Push-button switch
27. ____ Scan cycle
28. ____ Selector switch
29. ____ Setpoint (SP)
30. ____ Summing junction
31. ____ Temperature indicator
32. ____ Thermocouple
33. ____ Transmitter

A. The difference between the setpoint (the desired temperature) and the process variable (the actual temperature).
B. An electronic circuit that is mounted on an integrated circuit (IC) chip. The circuit receives digital numbers (in binary format) and converts the value to the equivalent voltage (0 to 10 volts) or current value (4–mA).
C. An electronic circuit that converts analog values such as 0 to 10 volts DC to digital values such as 0 to 4095.
D. An HVAC control circuit that is running in manual mode. For example, if the loop is controlling water flow to a fan coil, a valve is manually set or adjusted for the proper amount of flow to create the temperature that is desired.
E. A pneumatic or electric motor that is designed specifically for low rpm speed and usually makes only one revolution of its shaft to cause full movement of the damper.
F. A point in an automated process loop where the setpoint and process variable signal are compared.
G. An electronic device that allows you to convert the millivoltage signal from the thermocouple to a 0 to 10 VDC or 4–20 mA signal.

H. An acronym from "American Standard Code for Information Interchange." A standard for assigning numerical values (0 to 7) to the set of letters in the Roman alphabet and typographic characters used in computers.
I. An electronic system that provides control of an HVAC system.
J. A number that includes decimal values and/or exponential values. These types of numbers allow computers and electronic controllers to use very small decimal numbers or very large exponential numbers.
K. A numbering system that uses binary numbers "1" and "0" and place them into four weighted locations eight, four, two, and one. This numbering system represents decimal values 0 to 9 with the four binary numbers.
L. A base sixteen numbering system that consists of numerals 0 to 15. In electronic and PLC control systems the numbers are designed to take up only one column. This is accomplished by representing values 0 to 9 with traditional numbers and two-digit values 10 to 15 with letters of the alphabet, where 10 = A, 11 = B, 12 = C, 13 = D, 14 = E, and 15 = F.
M. A base eight numbering system that consists of values 0 to 7. This numbering system was used with early computer systems that had 8-bit functions.
N. A temperature sensor made of two dissimilar metals. When heat is applied to the tip of this sensor, it generates a small amount of DC millivoltage (1/1000 of a volt).
O. An electronic switch that can detect position similar to a limit switch. This switch uses an inductive field at the tip of the switch to detect the presence of a metal target.
P. An electrical switch that has an operator that allows the contacts to be closed or opened when the switch is activated. The contacts in this switch will remain in the position they are switched to until the switch is changed to a new position.
Q. A numbering system consisting of "1" and "0," primarily used in computers and programmable logic controllers (PLCs). Also called a base two numbering system.
R. A numbering system where the values are 0 to 9.
S. The input of this sensor is electrical current usually 4–20 mA. The transducer converts this current proportionally to air pressure, usually 3–15 psi.
T. A programmable digital electronic component that is mounted on an integrated circuit (IC). This device has a central processing unit (CPU) and it provides the function of a computer in handheld devices, instruments, and embedded systems such as programmable logic controllers (PLCs).
U. A module on a programmable logic controller (PLC) or a direct digital control (DDC) system that is designed to take a low-voltage signal from the controller and convert it to a full-voltage signal such as 12 VDC or 24 VDC or 115 VAC or 230 VAC for use in the system. The full-voltage signal can be used to turn on solenoids, motor starter coils, or other small loads.
V. A microprocessor-type controller that has input modules, output modules, and a controller. The controller has the sequential program saved in its memory and is executed continually as it checks switches that are connected to its input modules, and energizes or de-energizes output devices such as solenoids or motor starter coils that are connected to its output modules.
W. An electrical switch that is activated by motion. The switch senses linear or rotational motion, which causes its contacts to change state at some specified location or position.
X. A module in a programmable logic controller (PLC), where input switches are connected. Each switch is connected to an individual electronic circuit in this module, and this module sends a signal to the PLC to indicate if the switch is energized or de-energized.

Y. An electronic circuit that is mounted on an integrated circuit. The main function of this circuit is to provide amplification of an input signal. The op amp is used in a variety of electronic applications where it is combined with a sensor to increase the small signal that the sensor detects, and makes it large enough to be usable.

Z. The signal that comes from the sensor that provides feedback to the HVAC system. If the control system is controlling temperature, the signal would be the temperature that is measured by the temperature sensor.

AA. A switch that is activated by a person by depressing the switch.

BB. A temperature gauge or electronic display that shows the temperature of a system.

CC. A whole number that includes zero and can be a positive or negative value.

DD. A means of connecting two or more computers together so that information and data can be transmitted between them.

EE. The part of a programmable logic controller (PLC) that checks all inputs, executes the logic program, and then writes an "on" or "off" signal to the outputs in the output module.

FF. A device that senses pressure in an HVAC control system and converts it to electrical current. The typical air pressure is 3 to 15 pounds per square inch, and the current is 4–mA.

GG. The desired value or the target value of an automatic control system.

TRUE OR FALSE

Place a *T* or *F* in the blank to indicate if the statement is true or false.

Score __________

1. ____ The discrete inputs and outputs have two states: on and off.
2. ____ The octal numbering system uses only digits 0 to 7.
3. ____ The binary-coded decimal (BCD) system uses binary digits to display decimal numbers.
4. ____ An analog input has only two states: on and off.
5. ____ An isolated output module uses a relay to ensure total isolation.

MULTIPLE CHOICE

Circle the letter that represents the correct answer to each question.

Score __________

1. The range for the most typical analog voltage signals is:
   a. 4–20 mA.
   b. 0–10 V.
   c. 2–15 V.
   d. 0–110 V.

2. The range for the most typical analog current signals is:
   a. 4–20 mA.
   b. 0–10 V.
   c. 2–15 V.
   d. 0–110 V.

3. A 10-bit A/D converter provides _______________ resolution.
   a. one part in 256
   b. one part in 1024
   c. one part in 4096
   d. It is impossible to tell without more information.

4. The main difference between a thermocouple, RTD, and thermistor is that:
   a. a thermocouple produces a change in resistance, the RTD produces millivoltage, and the thermistor produces a millivoltage.
   b. a thermistor produces a change in resistance, the RTD produces millivoltage, and the thermocouple produces a millivoltage.
   c. an RTD produces a change in resistance, the thermistor produces millivoltage, and the thermocouple produces a millivoltage.
   d. an RTD produces a change in resistance, the thermistor produces a change in resistance, and the thermocouple produces a millivoltage.

5. A DDC and pneumatic system are:
   a. totally incompatible, and parts of the two systems should never be used together.
   b. compatible, and the pneumatic sensors and outputs can be connected directly into the DDC controller.
   c. compatible, and the pneumatic sensors must be connected to the DDC through P/I converters and outputs must be connected through I/P converters.
   d. compatible, and the pneumatic sensors must be connected to the DDC through I/P converters and outputs must be connected through P/I converters.

6. When the PLC processor is in the PROGRAM mode, it:
   a. executes its scan cycle.
   b. does not execute its scan cycle.
   c. is impossible to tell whether the processor is executing its scan cycle because you are not on line.

7. You can use the _______________ on the face of a PLC to determine whether an input switch is on or off.
   a. power light
   b. input status indicators
   c. output status indicators

LAB EXERCISE: OBSERVING AND UNDERSTANDING DIRECT DIGITAL CONTROL SYSTEMS AND PLC SYSTEMS

Safety for this Lab Exercise

In this exercise you will observe some systems that are controlled by direct digital controls (DDC) or by a programmable logic controller (PLC). You will primarily be observing, so you will have minimal risks for electrical shock hazards or risks to mechanical parts. As always, you must be aware of the operational system, and do not touch any exposed electrical terminals.

Tools and Materials Needed to Complete the Lab Exercise

Your instructor will provide one or more systems that are controlled by a DDC or PLC. You will be able to observe the operation of these systems as they control one or more parts of an air-conditioning or refrigeration system.

References to the Text

Refer to Chapter 20 in the textbook for additional information. You may need to read sections of the chapter again to help you understand the material in this exercise.

Sequence to Complete the Lab Task

Understanding a Closed-Loop System that Is Controlled by a Direct Digital Controller (DDC)

Figure 20–1 shows the basic diagram of a control loop. A loop is a control system that has a setpoint that sets the desired value for the system, a sensor that provides feedback, a controller, and an output device that turns on and off to control the variable for the system. When the loop is in automatic mode, the setpoint value is compared to the feedback signal from the sensor. If you place the loop in manual mode, an operator can control the output directly and set the amount of output signal that is sent to the output device from 0 to 100%.

The loop in Figure 20–1 is used to control a water valve to let cold water into a fan coil that has air blowing over it. The system uses a thermostat to determine if the conditioned space needs cool air or not. This system could also provide a second loop to control a warm water valve to provide heating for the system. In this diagram the thermostat provides two functions: First, it provides a setpoint for the system, and second, it provides the feedback function as a temperature sensor by providing the actual temperature of the system as a feedback signal. The temperature setting that you place the thermostat to is called the setpoint. For example, if you set the thermostat for 70°F, that is the setpoint for the system. If the room temperature is below 70°F, it will tell the system to add heat, and this is the feedback function of the thermostat. The feedback temperature is also called the process variable. The temperature setpoint is sent to the DDC as a number. The feedback temperature that the temperature sensor is measuring is also a number. The difference between the setpoint and the feedback temperature is called error. If the setpoint is warmer than the actual temperature the error signal is positive, and when it is sent to the controller, the controller sends a signal to the output amplifier, which sends a 0 to 100% signal to the water valve. The water valve accepts a 0 to 10 VDC signal that causes the valve to go from full closed (zero volts) to full open (10 volts). If the controller sends a 5-volt signal, the valve will open 50%.

Your instructor will provide a direct digital controlled (DDC) system. You should answer the following questions about this system. You will be asked to make changes to the setpoint and observe the systems operation if it is an operational system.

1. What is the output device for your system (what is the system controlling)? __________

2. What is the sensor for the system? __________

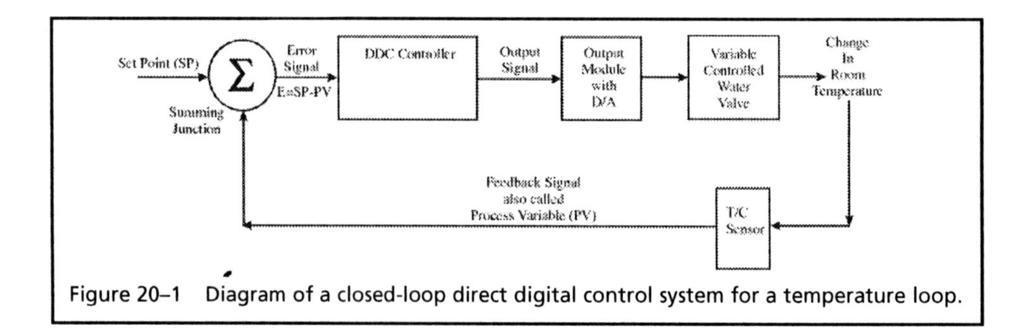

Figure 20–1 Diagram of a closed-loop direct digital control system for a temperature loop.

3. What is used to put a setpoint into your system? __________

4. What brand name of controller does your system use? __________

5. Explain in general terms what your system is controlling, and how it operates.

____

____

Sequence to Complete the Lab Task

Looking Closer at a Closed-Loop System that Is Controlled by a Direct Digital Controller (DDC)

Now that you have a general understanding of the DDC system, this section of the lab exercise will provide a closer look at the system. Figure 20–2 shows the same type of diagram as in Figure 20–1. This diagram shows the temperature sensor connected to an analog-to-digital (A/D) converter. The A/D converter is usually found inside an input module for the system. The analog signal from the temperature sensor is a voltage signal that ranges from 0 to 10 volts. The A/D converter changes the voltage signal to a digital number, which could be a value between 0 to 4095 if the system is using a 12-bit analog system. The digital number is used by the controller to compare the value coming from the temperature sensor to the value from the setpoint. The controller will determine if the output signal should increase or decrease from where it is. If the room is too warm, it will increase the signal to the cold-water valve to cause it to open more and allow more cold water to circulate through the coil. If the room is too cold, the controller will decrease the signal to the valve. If the temperature is at the setpoint, the controller will not change the output signal.

The controller sends a digital signal to the digital-to-analog (D/A) converter that converts the numerical value from the controller back to a voltage value (0 to 10 volts). The D/A converter is usually found inside an output module. The signal to the water valve is 0 to 10 volts. Some systems use a milliamp signal (4–0 mA) instead of a voltage signal. The milliamp signal uses an offset so that it does not start at zero. Since the lowest value is 4 mA, the system knows that it has a broken wire in the signal if the signal value goes to 0 mA. This allows the system to check for broken wires throughout the system by using the 4–20 mA system. Your instructor will show you the D/A and the A/D modules for your system. Answer the following questions about your system.

1. What is the voltage range (or current range) of your input sensor? __________

2. What is the voltage range (or current range) of your output sensor? __________

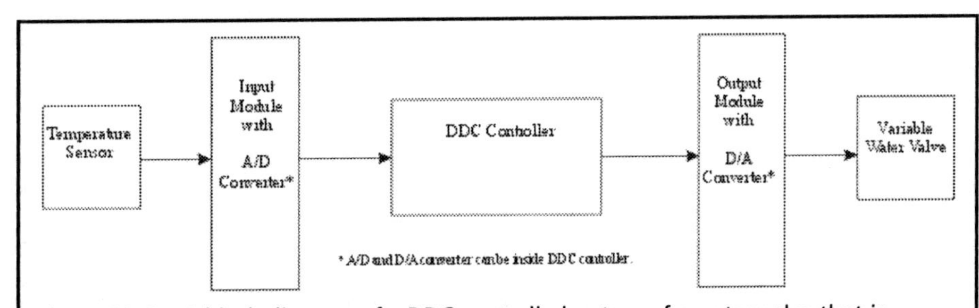

Figure 20–2 A block diagram of a DDC-controlled system of a water valve that is controlled by a temperature sensor. The temperature sensor has a setpoint and provides feedback to indicate how close the temperature is to the setpoint.

3. If the valve is set for 50%, how much voltage will be sent to it? __________

4. Explain how the information in questions 1 through 3 will help you troubleshoot a DDC system?

____

____

Sequence to Complete the Lab Task

Observing the Operation of Programmable Logic Controller

A programmable logic controller (PLC) is similar to a DDC system, except it uses an open architecture structure, which means you can use virtually any sensor and any output with the system. Some DDC systems are designed so that you must purchase the sensors and output devices from the system manufacturer, and only devices that they make will work with their system. This may become a problem when repair parts are not available, or the expense of one-of-a-kind devices make it difficult to repair older systems. The PLC system is designed to have any brand name of devices connected to it. This makes the PLC system very versatile.

The PLC can have on/off-type devices such as push-button switches, selector switches, or limit switches connected to its input modules and on/off devices such as solenoids, contactors, and motor starter coils. The PLC can also have analog input devices that have a voltage signal of 0 to 10 volts or 4–20 mA signals connected to its input module from a wide variety of temperature, pressure, flow, or position sensors. It can also have a wide variety of analog output devices such as motor drives, damper motors, and water valves that can accept voltage signals (0 to 10 volts) or current signals (4–20 mA). This allows a complete HVAC or refrigeration control system to be designed and controlled.

The PLC accepts a program called ladder logic that looks similar to the schematic (ladder) diagram. The programming software resides on a laptop or desktop computer and allows you to create a program or connect to an operating system and check the program that is running, and troubleshoot any input or output device. Figure 20–3 shows the basic part of the PLC. You can see input devices are connected to the input module, and output devices are connected to the output module. If the system uses analog sensors and outputs, they would be connected to an analog input module, and analog outputs would be connected to an analog output module. There are indicator lamps to show when an input device is sending voltage to the input module and when

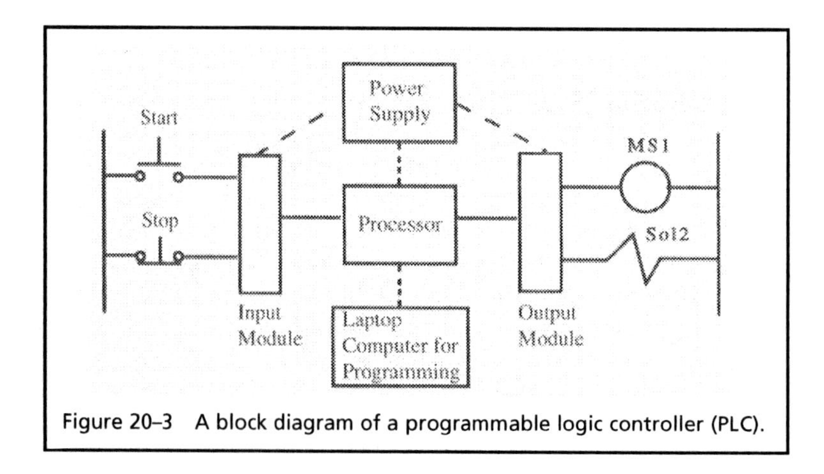

Figure 20–3 A block diagram of a programmable logic controller (PLC).

the PLC is sending a voltage to an output module. As a technician, you can check the input and output status lights to determine which inputs and outputs are energized. Figure 20–4 shows examples of these status lights. Your instructor will provide a PLC-controlled system, and you should observe its operation and answer the following questions.

1. What devices are connected to the PLC input module?

____

____

2. What devices are connected to the PLC output module?

____

____

3. Activate one of the input devices and observe the input status light. What do you notice about the status lights when the device is activated?

____

____

4. Have your instructor activate the input devices that will energize one of your output devices. What do you notice about the status light when the output device is activated?

____

____

5. What can you say about troubleshooting input and output devices that are connected to a PLC?

____

____

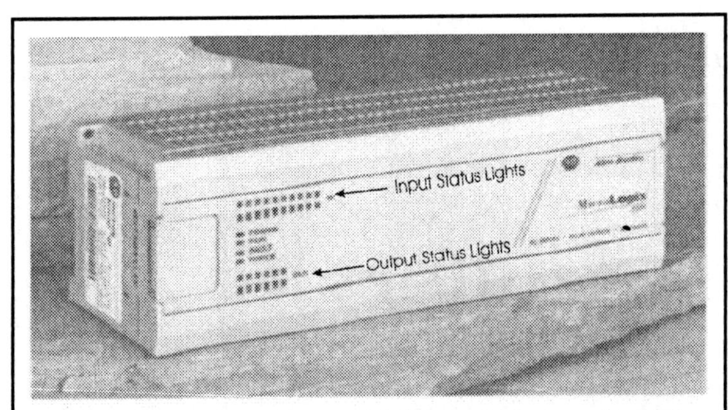

Figure 20–4 Example of the status indicators on input and output modules for the MicroLogix PLC. The input status lights are across the top, and the output status lights are shown across the bottom. *(Courtesy of Rockwell Automation)*

Checking Out

When you have completed this lab exercise, clean up your area, return all tools and supplies to their proper place, and check out with your instructor. Your instructor will initial here to indicate you are ready to check out. _______

CHAPTER 21

Getting and Keeping an HVAC and Refrigeration Job

OBJECTIVES

At the end of this lab exercise you will be able to:

1. Understand the function of a résumé and what goes into it.
2. Understand the function of a cover letter.
3. Create answers for questions for a job interview.

INTRODUCTION AND OVERVIEW

One of the most important parts of your education is the preparation for getting your first job and learning things to keep your boss happy. In this lab exercise you will create a résumé and cover letter for yourself and send it out to a number of prospective employers. You will learn how to identify job opportunities and send your résumé to them. You will also learn how to use the Internet to locate jobs and post your résumé. This lab exercise will help you understand what is expected of you when you go to a job interview and the types of questions you will be asked. When you have completed this lab exercise, you should feel comfortable about the process of applying for your first job or changing jobs. You will also have a good understanding of what your boss expects when you are hired.

TERMS

Cover letter
Goals
Job interview
Portfolio
Promotion
Résumé

MATCHING

Place the letter A–F for the definition from the list that matches with the terms that are numbered 1–6.

Score __________

1. ____ Cover letter
2. ____ Goals
3. ____ Job interview
4. ____ Portfolio
5. ____ Promotion
6. ____ Résumé

A. Examples of your work that show or demonstrate your knowledge and experience. This is an important part of putting together examples of your lab experience and other school experiences that you can show a prospective employer in preparation for your first job.
B. A list of things you would like to achieve in respect to your career.
C. A printed document that is created by an individual that outlines the person's education, job history, experience, and skills. The document is sent to a number of prospective employers when a person is looking for a job.

D. A letter that is written to a prospective employer that accompanies a résumé. This letter is used to introduce the person to the prospective employer and explain why he or she is applying for the job and what makes him or her the person the employer should invite to a job interview.
E. A meeting between an employer and the person who is seeking a job. This meeting occurs at the invitation of the employer and it is specifically designed for the person seeking the job to sell him- or herself to the employer and explain all of his or her skills and experiences.
F. A condition on the job where you get an increase in pay and take on additional responsibilities.

TRUE OR FALSE

Place a *T* or *F* in the blank to indicate if the statement is true or false.

Score __________

1. ____ You do not necessarily need a résumé if you just go ask somebody for a job.
2. ____ One of the most important things you can stress about a prior job that was not in the air-conditioning field, such as work in a fast-food restaurant, is your attendance record.
3. ____ A self-starter is someone who can be given a minimal amount of information about a job and be able to start the job without too much supervision.
4. ____ In the job interview the first thing you should discuss is the benefits, the pay, and how many days off you're going to get.
5. ____ You should get an invitation to a job interview for every single résumé that you send out.

MULTIPLE CHOICE

Circle the letter that represents the correct answer to each question.

Score __________

1. Your résumé should contain which of the following?
   a. Your name, address, and phone number
   b. Information about your skills
   c. Information about your previous jobs or work experience
   d. Information about your education
   e. All of the above
   f. Only b, c, and d

2. Your cover letter should contain which of the following?
   a. Your goals
   b. An introduction of yourself
   c. A short statement of why you want this job and why you think you are the best candidate for the job
   d. A complete list of all the courses that you have taken
   e. All the above
   f. Only a, b, and c

3. If you receive an invitation to a job interview you should:
   a. check out the company and gather as much information about it as you can.
   b. prepare a list of likely questions that you'll be asked and practice answering them.
   c. determine the location of your job interview so you can tell how long it will take you to get there so that you will not be late to the interview.
   d. prepare a portfolio showing your skills or the work that you've done in school.
   e. All the above
   f. Only a and b

4. Which of the following may keep you from getting a job offer after an interview?
   a. Having poor grades or low grade-point average
   b. Making a poor first impression, and dressing sloppily to the interview
   c. Having a poor attendance record at your previous job
   d. All the above
   e. Only a and c

5. Which of the following may get you dismissed or fired from your job?
   a. Being late to the job
   b. Calling off work too often
   c. Not being reliable
   d. Not getting along with other workers or not respecting your boss
   e. Not taking care of your appearance
   f. All the above
   g. Only a, b, and c

LAB EXERCISE: WRITING YOUR RÉSUMÉ AND COVER LETTER AND PREPARING FOR A JOB INTERVIEW

Safety for this Lab Exercise

In this lab exercise you will create a number of paper projects and will not be subjected to any safety hazards.

Tools and Materials Needed to Complete the Lab Exercise

Your instructor will help you develop a résumé and cover letter that you will send to a number of prospective employers. You will use the textbook to review the questions you will be asked at a job interview, and identify things that will cause you to be fired and things that will make your boss happy.

References to the Text

Refer to Chapter 21 in the textbook for additional information. You may need to read sections of the chapter again to help you understand the material in this exercise.

Sequence to Complete the Lab Task

Creating Your Résumé

When you are ready to apply for a job, you will be expected to send a résumé that indicates your job skills, education, and previous work history. The goal of your résumé is to impress a prospective employer to the point of inviting you to a job interview. The résumé is a piece of paper that represents you, and indicates why you are the best person to be hired for this job. In this exercise you will review the sample résumé in Figure 21–1. You will need to research the following information, which you will use to create your personal résumé.

1. What is your personal information?
   a. Name

   ____

   b. Address

   ____

Jerry Smith
201 East State Street
Anytown, Ohio 45678
555-255-2001
Jsmith@atown.net

Career Objective
A position in an HVAC company where I can use my technical skills to install and troubleshoot residential and commercial HVAC equipment.

Education
2007 Associate's Degree in Applied Science; Major in Heating, Ventilating, and Air Conditioning
Terra Community College, Fremont, Ohio

2006 Certificate in HVAC Installation
Terra Community College, Fremont, Ohio

Skills
I have acquired the following skills and knowledge during my training:
- Using a volt-, ohm-, and milliammeter to troubleshoot
- Reading and creating electrical schematic (ladder) diagrams and wiring diagrams to install and troubleshoot HVAC equipment
- Complete understanding of the operation of all electrical controls for HVAC equipment
- Installation skills for furnaces, air conditioners, and heat pumps
- Complete knowledge of troubleshooting single-phase and three-phase motors and compressors
- Excellent customer service skills

Prior Work Experience
1998–2002 Grill Cook, Joe's Restaurant
Perfect attendance and excellent customer service skills.

2003–2006 Electrical Department, City Hardware
Duties included sales knowledge of all electrical components. Helped customers select proper electrical components.

2006–2007 HVAC Installation Technician, Northwest Heating and Cooling
Duties included installation of furnaces and air conditioners.

References are available upon request.

Figure 21–1 A sample résumé.

c. Phone number

____

d. E-mail address, if you have one

____

2. What is your career objective? What job are you interested in doing?

____

____

3. What is your education history?

a. High School

Where ____________________ Date graduated ____________

b. College

Where ____________________ Dates ____________

Types of courses ______________________________

Certificate ______________________________

Degree ______________________________

Other ______________________________

4. List the skills that you have learned that would apply to the career objective.

a. ______________________________

b. ______________________________

c. ______________________________

d. ______________________________

e. ______________________________

f. ______________________________

5. List your prior work history.

a. Where ______________________________

Dates ______________________________

What skills did you learn? ______________________________

b. Where ______________________________

Dates ______________________________

What skills did you learn? ______________________________

c. Where ______________________________

Dates ______________________________

What skills did you learn? ______________________________

When you have all of the information in 1 through 5 completed, you can put it into a document on a computer. This document will become your résumé and you will make a number of copies to send to prospective employers. You need to put this information into a computer system so you can easily make changes and print as many copies as needed. It is also important to have several people look over your résumé to check for spelling errors and other problems. Enter your résumé into the computer and print copies for several people to check. (*Note:* Your instructor may have you request help from an English instructor or other teachers to help review your résumé. They may also suggest a different format for your résumé that will work better in your local area.)

Sequence to Complete the Lab Task

Creating Your Cover Letter

When you are ready to apply for a job, you will be expected to send a cover letter to accompany your résumé. The cover letter will introduce you to the prospective employer. It is your chance to present a "first impression" so that the employer will be so impressed with you that you will be offered a job interview where you can meet the employer face to face. It is very important that the cover letter provides all the details about you, but it must be short and concise. In this exercise you will review the sample cover letter in Figure 21–2. You will need to research the following information, which you will use to create your personal cover letter.

1. Research the newspapers in your area or on the Internet to find a job that is advertised. Write a cover letter for this position.
   a. Name of the company ____________________
   b. Address ____________________
   c. City ____________________
   d. State ____________________
   e. Zip code ____________________

2. What is the job that is advertised?

   ____

3. List the skills that you have learned that you will put into the cover letter.
   a. ____________________
   b. ____________________
   c. ____________________
   d. ____________________
   e. ____________________
   f. ____________________

4. When you have all of the information in 1 through 3 completed, you can put it into a document on a computer. This document will become your cover letter and you will make a number of copies to send to prospective employers. You need to put this information into a computer system so you can easily make changes and print as many copies as needed.

Jerry Smith
201 East State Street
Anytown, Ohio 45678
555-255-2001
Jsmith@atown.net

May 23, 20XX

Mr. James White, Owner
Commercial Heating and Cooling
2678 Main Street
Anytown, Ohio 45678

Dear Mr. White,

I am responding to your advertisement in the local newspaper last week for an HVAC technician to install and troubleshoot furnaces and air conditioners. I am enclosing my résumé for this position with your company. I have just completed an Associate's Degree in Applied Science with a major in HVAC. During that time, I learned many skills that have prepared me for this position. My résumé outlines these skills and experiences. I have gained additional technical skills and customer service skills from my current and previous jobs. I have a proven track record of working with others as a team member and I have had an excellent attendance record on these jobs.

I think you will see from the résumé that I can use a volt-, ohm-, and milliammeter to troubleshoot a variety of electrical controls and motors. I have excellent skill in reading and creating electrical schematic (ladder) diagrams and wiring diagrams to install and troubleshoot HVAC equipment. I have a complete understanding of the operation, installation, and troubleshooting of all electrical controls for HVAC equipment. I have two years' experience in installing furnaces, air conditioners, and heat pumps. I have learned the skills used to troubleshoot single-phase and three-phase motors and compressors. I am always eager to continue learning the skills and acquiring knowledge to improve myself.

Thank you for your consideration. Please do not hesitate to contact me at the address or phone number provided above if you have any questions.

Sincerely,

Jerry Smith

Figure 21–2 Sample cover letter.

It is important to understand that you will need to change the cover letter so that it matches each job you are applying for. You should make sure that the cover letter is on a computer so you can change the name and address of the company that is placing the job add.

It is also important to have several people look over your cover letter to check for spelling errors and other problems. Enter your cover letter into the computer and print copies for several people to check. (*Note:* your instructor may have you request help from an English instructor or other teachers to help review your cover letter. The may also suggest a different format for your cover letter that will work better in your local area.) Save this cover letter so that you can send it to jobs that are listed in your area.

Sequence to Complete the Lab Task

Creating Answers to Typical Questions You Will Be Asked in a Job Interview

If your résumé and cover letter has impressed a prospective employer, you will be invited to a job interview. During the job interview the employer will ask a number of questions to learn whether you are the person they will eventually hire.

The next section of this lab will show some questions you might be asked on a job interview. Create answers for each of the questions listed below. Your instructor will help you select a company in your area to use for this exercise. You can pretend that this company has invited you to a job interview.

1. Tell me how your previous jobs or experiences have prepared you for this position.

   ____

   ____

2. Can you give me an example of where you have used teamwork to get a task completed?

   ____

   ____

3. Can you provide an example of where you used customer service skills to help your company work out a problem with a customer? (If you have not had a previous job, how would you deal with a problem with a customer?)

   ____

   ____

4. What motivates you?

   ____

   ____

5. How do you get along with co-workers?

   ____

   ____

6. Describe a difficult situation you have overcome.

   ____

   ____

7. Why are you the best person for this job?

   ____

   ____

8. What interests you about this job?

____

____

9. What do you know about this company?

____

____

10. What can you contribute to this company?

____

____

11. Is there anything about this company you would like to know more about?

____

____

12. What are your goals for the next five years?

____

____

13. What are your goals for the next 10 years?

____

____

14. What are your salary expectations now, and in five years?

____

____

15. What type of opportunities in this company would excite you?

____

____

16. Could you tell me why you think job attendance is important for this job?

____

____

17. Can you give me an example of how you have exhibited exceptional honesty or integrity?

____

____

Checking Out

When you have completed this lab exercise, clean up your area, return all tools and supplies to their proper place, and check out with your instructor. Your instructor will initial here to indicate you are ready to check out. _______